The
Adaptive Seascape

The
Adaptive Seascape

The Mechanism of Evolution

David J. Merrell

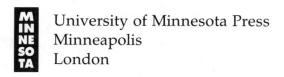

University of Minnesota Press
Minneapolis
London

Published by the University of Minnesota Press
2037 University Avenue Southeast, Minneapolis, MN 55455-3092

Library of Congress Cataloging-in-Publication Data

Merrell, David J.
    The adaptive seascape : the mechanism of evolution / David J. Merrell.
       p.   cm.
    Includes bibliographical references and index.
    ISBN 0-8166-2348-1 (acid-free)
    1. Evolution (Biology)   I. Title.
    QH366.2.M47   1994
    575—dc20   93-29841
                                                                      CIP

The University of Minnesota is an
equal-opportunity educator and employer.

Contents

of Behavior—Selection and Behavior—The Role of the
Environment—The Evolution of the Eye—Regressive
Evolution—The Biological Environment—Approaches
to the Study of Evolution

Preface

This book is an exposition and critique of the modern synthetic theory of evolution. The style of presentation is designed to make the ideas accessible to the widest possible audience, but the book is also suitable for use in advanced courses and seminars in ecological genetics, population genetics, and evolutionary biology.

A primary purpose is to deal with some of the current problems and controversies in evolutionary genetics. That evolution is gradual has almost become dogma. The case is made here for rapid saltational evolution as well as gradual evolution. Furthermore, evolutionary change is widely believed to result from the accumulation of numerous genes of small effect, but here the importance of genes of major effect, especially dominant genes, as well as polygenes in bringing about evolutionary change is emphasized. These concepts are used to interpret the evolution of industrial melanism, DDT resistance, and mimicry.

In the treatment of the nature and origin of species, an adaptive seascape is suggested as a more appropriate metaphor for Sewall Wright's well-known "adaptive landscape" because, with the physical and biological environment constantly changing, the adaptive surface more closely resembles that of the sea than that of the land.

The concluding chapters deal with some of the more controversial aspects of evolutionary theory. Included are discussions of neutralism, the molecular clock, molecular versus morphological evolution, group selection, and cultural evolution. The supposed evolutionary paradoxes related to the costs of sex, males, meiosis, variability, and even of natural selection itself are shown to be inherent in the underlying assumptions. With different, more valid assumptions, the paradoxes vanish. In the chapter on behavior and evolution, the influence of the biological environment and behavior on the course of evolution is discussed.

Many of the ideas presented here were first tested in a seminar on ecological genetics that has been offered since 1968 in collaboration with Professors William D. Schmid and James C. Underhill of the University of Minnesota. I am grateful to them and to the many graduate students who participated for their interest and support.

Chapter 1
The Development of Modern
Evolutionary Theory

Modern evolutionary theory, also known as the modern synthesis, has in recent years become the subject of criticism, with the critics usually concluding that the modern synthesis is outmoded and that new paradigms are needed. Further reading too often reveals, however, that the critics' understanding of the modern synthesis has never gotten past Darwin, and that they equate the modern synthesis with natural selection. For those who regard Darwinian natural selection as the essence of the mechanism of evolution, apparently none of the advances in evolutionary theory since 1900 seems particularly significant.

On the other hand, there is another group of critics whose understanding of the modern synthesis appears to be based on Ernst Mayr's 1963 work, *Animal Species and Evolution*, and who seem unfamiliar with much of the work prior to that time. Although Mayr's book has been very influential, two of the major architects of the modern synthesis wrote scathing reviews of his lack of understanding of their work. Sewall Wright (1960) in his typical methodical way made a point-by-point response to Mayr's 1959 paper, "Where Are We?", showing how Mayr had misunderstood the significance of his work in population genetics. J. B. S. Haldane (1964), in "A Defense of Bean-bag Genetics," was even more critical of Mayr's interpretation of his work. Since these papers by Wright and Haldane appeared, the history of the development of the modern synthesis has drawn increased attention, notably in *The Evolutionary Synthesis* (1980), edited by Mayr and William Provine.

This chapter presents my own perception of the significant events that led to the formulation of the modern synthetic theory of evolu-

1

tion and will serve as a prelude to the exposition and critique of the modern synthesis that follows.

Darwin

Although the idea of evolution had cropped up many times, only with the publication in 1859 of Charles Darwin's *On the Origin of Species by means of Natural Selection* did the idea win widespread acceptance among scientists. Darwin is also credited with suggesting the first plausible mechanism for mediating evolutionary change, natural selection. Although Alfred Russel Wallace deserves credit as the co-discoverer of natural selection, and Darwin has sometimes been accused of plagiarizing this idea from Wallace and others, the cordial relations between Darwin and Wallace throughout their lives suggest that this accusation has little merit.

Darwin's theory of natural selection was so simple and straightforward that Thomas Henry Huxley, Darwin's strongest advocate, chastised himself for not thinking of it first. It consists of certain observations and the conclusions to be drawn from them. The first observation is that the reproductive capacity of all species is geometrical, far greater than necessary to maintain the existing numbers of the species. Darwin's calculations showed this to be true even for elephants, which he took to be the slowest breeding of all species. Moreover, under extremely favorable conditions, some species, such as the rabbit introduced into Australia, have demonstrated that a dramatic, rapid increase in numbers actually does occur.

The second observation is that, despite this enormous reproductive capacity, in most species there is no geometrical increase in numbers. Instead, the population size remains more or less constant, fluctuating about some average number. These two observations led to Darwin's first conclusion: only a fraction of the young produced by a species ever reaches sexual maturity. For example, leopard frog females produce an average of about 3,000 fertilized eggs per female. When the population size remains constant, however, only two offspring per female on average must survive to sexual maturity.

Darwin's third observation was that no two individual members of a species (I might add, not even identical twins) are ever completely identical; biological variation is universal.

These observations led to Darwin's final conclusion: some individuals in a species will be better able (or better adapted) to survive and reproduce under the existing physical and biological conditions than others. Even though many of the deaths may be random, and much of the variation may be environmental, if the differences between the individuals that live and those that die are to any extent hereditary, these differences will be reflected in the hereditary composition of the next generation, for the next generation will have a greater proportion of the favored hereditary variations than the previous generation. This process, which Darwin called natural selection, could, by operating over a number of generations, gradually change the characteristics of the species.

Darwin recognized that the greatest weakness in the theory was his lack of knowledge about the inheritance of variation. For this reason he flirted with the Lamarckian concept of the inheritance of acquired characteristics and also proposed his own theory of heredity, known as pangenesis, but unfortunately both theories were wrong.

Mendelism

The first real progress toward understanding heredity came with the rediscovery in 1900 of Gregor Mendel's laws on the segregation and independent assortment of hereditary factors, later called genes. In the late 1800s cytology had come into its own as the details of cell division and gametogenesis and of chromosome behavior during mitosis and meiosis were worked out. In this way the groundwork was laid for the development of the chromosome theory of heredity by Thomas Hunt Morgan and his associates, who established the principles of linkage and the linear order of the genes. Their work established not only the physical basis of heredity but also the physical basis of evolution, for it was clear that evolutionary change had to occur within the physical framework imposed by the chromosomes.

Mendelian inheritance was particulate, and initially geneticists dealt with easily classified qualitative traits. However, Darwin's cousin, Sir Francis Galton, had started to study the inheritance of quantitative traits prior to 1900. These studies did not lead to any striking breakthroughs in the understanding of heredity, but the methodologies developed became the foundation for modern statistics. Galton was succeeded in these efforts by Karl Pearson and then by R. A. Fisher.

Also prior to 1900, William Bateson had published *Materials for the Study of Variation*, in which he concentrated on discontinuous or qualitative variations as the materials for evolution. Thus he was mentally prepared for and very receptive to Mendel's ideas and became the foremost proponent of Mendelism in Great Britain.

A bitter dispute developed between Bateson, on the one hand, and the biometricians, followers of Galton, on the other, who felt that Mendelizing qualitative traits were superficial and trivial and that the more important quantitative traits followed different rules of inheritance. The biometricians argued that a particulate theory of inheritance, Mendelism, could explain the inheritance of discrete qualitative traits, but not the inheritance of continuously varying quantitative traits.

The Dutch botanist Hugo De Vries, a codiscoverer of Mendel's papers in 1900, was also receptive to Mendelism, for in his own experiments with plant hybrids, he had obtained results like Mendel's even before he had read Mendel's work. Within a very short time (1901–3), De Vries had proposed the mutation theory for the origin of species as an alternative to natural selection. He was led to this theory by his observations on the evening primrose, *Oenothera lamarckiana*, in which he found new true-breeding varieties arising at a single step that behaved like new species. Thus, the mutation theory postulated a saltational origin for new species rather than the gradual origin envisioned with natural selection.

In the early years of this century, the science of genetics developed explosively; it was the hot new field where all the action was, very reminiscent of the development of molecular biology in recent years. It is of considerable interest, therefore, that three of the leading geneticists in the world, De Vries on the Continent, Bateson in Great Britain, and Thomas Hunt Morgan in the United States, were skeptical about the theory of evolution by natural selection and sought other explanations for the origin of species.

One cannot help but wonder if their thinking was not influenced by the kind of experimental material they worked with. This was certainly true of De Vries, but in the early days of genetics, nearly all geneticists sought out simply inherited, clear-cut, qualitative traits in order to work out the laws of heredity, and their numerous successes may have led them to a certain arrogance toward such old-fashioned ideas as natural selection. Furthermore, a mutational, saltational origin of species was a lot simpler to understand than the more gradual process envisioned with natural selection. Given the widespread acceptance of natural selection today, it may be difficult to appreciate how low its status became in the early years of the century, but a quotation from Nordenskiöld's well-known *History of Biology* (1920–24; 1928) reflects the attitudes of that time: "To raise the theory of selection, as has often been done, to the rank of a 'natural law' comparable in value with the law of gravity established by Newton is, of course, quite irrational, as time has already shown; Darwin's theory of the origin of species was long ago abandoned. Other facts established by Darwin are all of second-rate value."

Although De Vries's "mutations" giving rise to new species of *Oenothera* turned out to be chromosome recombinants for the most part, rather than true mutations in the usual sense, his mutation theory nevertheless called attention to a new process, independent of natural selection, by which the hereditary material of a species could change. Thus originated the second component part of the modern synthesis.

Johannsen

Soon after the rediscovery of Mendelism, Wilhelm Johannsen, a Danish botanist, carried out some experiments with garden beans of fundamental importance to genetics and to the study of evolution. These beans are normally self-pollinating so that over a number of generations they become highly inbred. Any plant resulting from continuous self-pollination will be homozygous at virtually every locus, and its descendants in the F_1 and F_2 generations will all have genotypes similar to one another and to the original plant. Such a line of descent is called a pure line. Johannsen's experimental material consisted of nineteen different, independently inbred, pure lines of beans.

Since the significance of Johannsen's results (1903, 1909) has sometimes been misinterpreted (e.g., Mayr, 1970, p. 181), it is worth outlining his results. When Johannsen weighed the individual beans, he found a wide range of variation in weight within lines as well as between lines. The average weights for the different lines were significantly different, however. Furthermore, when progeny from high-weight and low-weight lines were grown, the seeds on plants from a high-weight line weighed significantly more, on average, than those on plants from a low-weight line. Therefore, the different lines must have been genetically different for seed weight, and selection between lines for high or low weight was effective.

As noted, there was a wide range of variation in seed weight within a given line and even among the beans on a single plant. For a number of generations Johannsen selected the largest and smallest beans within a single line and reared their progeny, trying to establish large and small sublines within the original line. Even though these beans were sometimes strikingly different in size, the average weight of their progeny was the same. Thus, selection within a line was futile because the inbred line had become homozygous and lacked genetic variation. All of the variability seen within an inbred line was environmental, owing to the number and position of the beans in the pod, the number of pods, the time of seed set, and so on.

Johannsen's results showed for the first time that a given trait exhibits both hereditary and environmental variation, and that the best way to determine the magnitude and direction of these effects is through experimental breeding tests. Moreover, the environmental effects are great enough to mask possible genetic discontinuities so that seed weight is a continuously varying, quantitative trait.

Darwin had never made a clear-cut distinction between hereditary and environmental variation, as evidenced by his willingness to accept Lamarckism, the theory that acquired characters are inherited. He apparently felt that selection could be effective on all variation. Johannsen showed, unequivocally, that selection on environmental variation was futile; only if genetic variation was present was there a response to selection. Once the hereditary variation was exhausted, selection was no longer effective, for it could not create new variation.

These results also led Johannsen to coin the words and draw the crucial distinction between the genotype (the total hereditary constitution of an individual) and the phenotype (the sum total of all the traits manifested by an individual and produced by its genotype interacting with the environment).

Population Genetics

Experimental genetics routinely requires the use of controlled matings, say of type A with type B, so that meaningful data can be obtained. The question arose very early, however, about what happens in nature where matings are not controlled. William Castle, in a 1903 footnote, was the first to address this question, followed by Karl Pearson in 1904, and by G. H. Hardy and W. Weinberg independently and in a more complete fashion in 1908. The simple but fundamental rule that emerged has gone under various names, the most appropriate being the Castle-Hardy-Weinberg (C-H-W) law. The formulation was mathematical, but the math was so simple that Hardy, a mathematician, reported his findings in a brief letter to *Science* rather than in a more formal paper. Simply put, it was shown that, in a randomly mating large population in which it is assumed that there is no mutation, migration, or natural selection, the frequencies of the genes in that population will remain constant and unchanged from one generation to the next. Moreover, after one generation of random mating the genotypes will be in specified proportions and will remain so as long as random mating continues. At first glance, this result hardly seems earth-shaking. However, when it is recalled that Darwin believed in blending inheritance and thought that variability was lost every generation and had to be replenished, then it will be realized that the C-H-W law removed a major stumbling block from Darwin's thinking at one fell swoop and in the simplest possible manner. No longer was it necessary to find a mechanism to explain how large amounts of genetic variability were replenished each generation; the variability was never lost.

Another conclusion from the law is that evolutionary change does not occur automatically. The population will remain unchanged unless certain identifiable factors are at work. These factors can be recognized in the assumptions underlying the C-H-W law. The assumption of a large population size is necessary because in small populations random genetic drift owing to accidents of sampling may lead to changes in gene frequency. Random mating is assumed because nonrandom mating (selective mating, assortative mating, or inbreeding) will lead either to gene frequency changes or at least to deviations from the expected C-H-W equilibrium frequencies. Although mutation rates are ordinarily so low that their effects on gene frequency change in the short run may be trivial, their cumulative effects over time cannot be ignored. Hence the assumption of no mutation is made. If new genes are introduced into a population by immigrating individuals, obviously its gene pool will be enhanced and in all likelihood changed in composition; and if individuals with certain genotypes survive and reproduce better than individuals with other genotypes, the gene pool also will change. Hence the ban on immigration and selection was imposed. Admittedly, to expect all of these assumptions to hold in any population is unrealistic. This hypothetical population is the starting point, however, for our understanding of how the evolutionary mechanism works and how evolution must occur.

The controversy between the Mendelians and the biometricians over whether a Mendelian explanation was possible for the inheritance of quantitative traits continued until about 1910 when H. Nilsson-Ehle (1909) and E. M. East (1910) independently formulated the multiple factor hypothesis, a Mendelian explanation for the inheritance of quantitative traits. This theory postulated that quantitative traits are influenced by genes at a number of different loci, each of small effect, as well as by environmental influences. Although alleles at each locus segregate in Mendelian fashion, when many such loci with similar and cumulative effects segregate, a large number of different genotypic classes for the trait will be formed, each slightly different from the next. Moreover, the different classes will form a normal frequency curve, with the intermediate phenotypic classes the most frequent. The differences between adjacent genotypic classes will be so slight that they are difficult to distinguish, and the slight genetic discontinuities will be blurred by environmental effects to the point where the phenotypes generated by the different genotypes overlap and the variation is continuous and quantitative. The multiple factor hypothesis assumed genes at a number of loci with similar small additive effects, without dominance, and independently inherited. Although these assumptions seem too simple, the hypothesis held up remarkably well in the interpretation of data.

The controversy over quantitative inheritance was finally ended by R. A. Fisher's classic paper (1918), "The Correlation between Relatives on the Supposition of Mendelian Inheritance." Fisher, originally trained in mathematics and biometry, became one of the founders not only of modern statistics but of population genetics as well. In this paper, he took into account the possibility that the numerous genes affecting quantitative traits might differ in the magnitude of their effects, in their degree of dominance, and in their linkage relations. He also considered the possible effects of multiple allelism, epistasis, and the environment. This paper reconciled the two schools of thought, Mendelian and biometrical, and became the basis for subsequent developments in the understanding of quantitative inheritance.

By this time the stage was set for the development of theoretical or mathematical population genetics, which was primarily the result of the efforts of four scientists, S. S. Chetverikov, Fisher, Haldane, and Wright. This work became the foundation for the modern synthetic theory of evolution. There have been some efforts to downplay the contribution of theoretical population genetics to the development of the modern synthesis. For instance, in the first chapter of *Evolution* (1977), a textbook by Dobzhansky, Ayala, Stebbins, and Valentine, Stebbins wrote (p. 17), "The modern synthetic theory as a generally accepted way of approaching problems of evolution was born in 1937 with the publication of Dobzhansky's *Genetics and the Origin of Species*." In the fifth chapter of the same book, however, Dobzhansky himself wrote (p. 129): "The present book is essentially an exposition of the modern theory of evolution, called the 'biological' or 'synthetic' theory of evolution. The fundamentals of this theory were arrived at, largely independently, by Chetverikov, Fisher, Haldane, and Wright between 1926 and 1932; but the theory was developed by many authors, starting in the 1930s." In his final work on the modern synthesis, published in 1970 and called *Genetics of the Evolutionary Process*, Dobzhansky also referred to "the creators of the modern biological theory of evolution, S. Tshetverikov (1926), R. A. Fisher (1930), J. B. S. Haldane (1932), and Sewall Wright (1931)." Although such notables as Ernst Mayr (Mayr and Provine, 1980, p. 421), Hampton Carson (Mayr and Provine, 1980, p. 91), and G. Ledyard Stebbins attribute the start of the modern synthesis to Dobzhansky's 1937 work, I can only agree with William Provine (1978), who wrote that those evolutionists who claimed not to have been influenced by the theoretical geneticists, but rather by Dobzhansky, did not recognize his indebtedness, and theirs, to Chetverikov, Fisher, Haldane, and Wright because they had never read their work. Thus, although others may have first become aware of the modern synthesis through the writings of Dobzhansky, he himself was well aware of his debt to his predecessors.

Furthermore, anyone familiar with the writings of these authors can easily recognize their influence in the three editions of *Genetics and the Origin of Species* (1937, 1941, 1951). Chetverikov's work was unknown to most biologists because it was in Russian, but when a translation was published in 1961, his influence on Dobzhansky became obvious, especially in the 1937 edition. The work of Fisher, Haldane, and Wright, on the other hand, was probably unfamiliar to biologists because in the 1930s and 1940s biologists were thought not to need much mathematics, and most of them were poorly trained in the subject. Dobzhansky had such a great impact because he had a dynamic personality and was a skilled writer with the wit and wisdom to turn the rather dry, abstract, mathematical works of the big four into exciting prose that was understandable and accessible to the great majority of biologists. However, if one reads Fisher's *Genetical Theory of Natural Selection* (1930) and Haldane's *The Causes of Evolution* (1932), as well as some of Sewall Wright's papers, before reading Dobzhansky, the derivation of his ideas is clear. Wright's work was the least accessible, but C. C. Li's *An Introduction to Population Genetics*, published in Peiping in 1948, made Wright's ideas available to a wider audience.

Population genetics provided the framework, the structure, on which the modern synthesis was constructed. The population geneticists showed that evolutionary change resulted from the effects of one or more of the four factors, namely, mutation, selection, migration, and random genetic drift. Each of these alone could bring about a shift in the prevailing C-H-W equilibrium, and thus, by definition, an evolutionary change. In combination, they provided a mechanism of evolution with both random and directional components that provided a useful model on which to base observations and experiments. Without the insights derived from population genetics, the modern synthesis would not have developed so rapidly or with such impact in so many fields.

The Modern Synthesis

Dobzhansky must be considered a disciple of the pioneers in population genetics, and the evidence is now quite clear from Provine (1986) how dependent he was on Sewall Wright for both theoretical and experimental ideas. Similarly, it is of considerable interest that the three most frequently cited authors in Julian Huxley's *Evolution: The Modern Synthesis* (1943) are Haldane, Wright, and Fisher, even though Huxley made no effort to discuss evolution in mathematical terms. These two books, *Genetics and the Origin of Species* and *Evolution: The Modern Synthesis*, are widely regarded as landmarks in the development of the synthetic theory of evolution, and in both cases their dependence on the developments in population genetics is unmistakable.

Another influential book in the development of the synthesis was Mayr's *Systematics and the Origin of Species* (1942) in which he showed how modern evolutionary theory could be applied to systematics. The source of Mayr's understanding of the genetics of populations is clear from his citations; thirty-eight from Dobzhansky, but only fifteen from Wright, three from Fisher, and none at all from Haldane. Stebbins, in *Variation and Evolution in Plants* (1950), introduced the concepts of the modern synthesis to botanists, but he too must have drawn his primary inspiration from Dobzhansky, for there were forty-six references to Dobzhansky, but only five each from Fisher and Haldane and sixteen from Wright. Simpson, in *Tempo and Mode in Evolution* (1944), brought the modern synthesis to paleontology. His citations are relatively few, but almost equally distributed among Dobzhansky, Fisher, Haldane, and Wright. Furthermore, in the preface to *The Major Features of Evolution* (1953), he wrote with respect to the writing of *Tempo and Mode in Evolution*, "At that time it was to me a new and exciting idea to try to apply population genetics to interpretation of the fossil record and conversely to check the broader validity of genetical theory and to extend its field by means of the fossil record." Thus, Simpson clearly acknowledged that his inspiration came from population genetics.

The role of Dobzhansky, Mayr, Simpson, and Stebbins was that of apostles or disciples, witting or unwitting, who applied the new evolutionary concepts developed by the population geneticists to their own disciplines. Another book of considerable influence was *The New Systematics* (1940), edited by Huxley, which included articles by authors with a wide range of interests. Bernhard Rensch's *Evolution above the Species Level* (1960) and I. I. Schmalhausen's *Factors of Evolution* (1949) are often included among the key elements in the development of the modern synthesis, but they had less impact than the books cited previously. M. J. D. White wrote a comprehensive account of animal cytogenetics in *Animal Cytology and Evolution* (1945), while C. D. Darlington had earlier produced a similar classical work on plant cytogenetics entitled *Recent Advances in Cytology* (1937) and a significant book, *The Evolution of Genetic Systems* (1939). Edgar Anderson's little book, *Introgressive Hybridization* (1949), aroused considerable interest when it appeared.

In the Soviet Union, following the lead of Chetverikov, a number of workers became interested in evolutionary problems and the genetics of natural populations, among them Dobzhansky, Dubinin, Timoféeff-Ressovsky, Schmalhausen, and Vavilov. It is interesting to speculate how far this work might have gotten if genetics had not been attacked by the Lysenkoists under Stalin. Dobzhansky and Timoféeff-Ressovsky emigrated; the work of Dubinin and Schmalhausen was interrupted, and they were forced to work in other disciplines. Vavilov was imprisoned and died.

In France, L'Héritier and Teissier were among the first to try to test some of the mathematical concepts of population genetics in experimental populations of *Drosophila* (1934). Malécot contributed to the mathematical theory on population structure, and Lamotte studied population structure and drift in natural populations of snails, but for the most part, the modern synthesis did not win much interest in France.

The situation in Great Britain was rather different. There, the influence of Fisher and Haldane was pervasive. One result was Huxley's book; another was the collaboration that developed between Fisher and E. B. Ford in which Ford attempted to study the genetics of natural populations within the framework laid down by the population geneticists. His approach was strongly selectionist and adaptationist, following Fisher's lead, and resulted eventually in the bitter controversy between Fisher and Ford on the one hand and Sewall Wright on the other over the evolutionary importance of genetic drift. Ford also became one of the first to work in the field that has come to be known as ecological genetics.

Another outstanding worker in ecological genetics was P. M. Sheppard, who studied natural populations of butterflies and snails. In 1954 Sheppard wrote: "The great advances in understanding the process of evolution, made during the last thirty years, have been a direct result of the mathematical approach to the problem adopted by R. A. Fisher, J. B. S. Haldane, S. Wright, and others. However, mathematical theories give no information on the conditions found in nature, but only show in what circumstances different factors can be of importance in evolution. . . . The mathematical treatment of evolution makes it clear that mutation rates, as low as those found in the laboratory, can only be important in controlling the direction of evolution in exceptional circumstances, but that isolation, population size, genetic drift and natural selection can all be important factors in bisexually reproducing organisms, although their actual importance in nature is a matter of dispute. The controversy will be only resolved by field work and not by mathematical argument. Consequently, the hypotheses derived by mathematicians have given a great impetus to experimental work on the genetics of populations." (See also Sheppard, 1953.) Thus, Sheppard, like Dobzhansky, gave credit to the population geneticists for the advances beyond Darwinism in our understanding of the evolutionary process. He also recognized the limitations of the mathematical approach—namely, that it could elucidate the possibilities under which evolutionary factors could bring about evolutionary change, but that definitive answers as to which factors were actually important in any given situation could not be determined by mathematical argument but required either field studies or experimental work (or both) on the genetics of the populations in question before the questions could be resolved. The attacks mounted by C. A. Waddington (1953, 1957) and Mayr (1959) on the value of the mathematical approach to the study of evolution seem not to have recognized its limitations or for that matter its inspirational effect on so many students of evolution. Its value lay not so much in answering questions as in raising them, in giving form and direction to evolutionary research.

There were two major weaknesses in the mathematical approach. The first was that most of the math dealt with one or two loci, obviously a gross oversimplification, although Haldane, Fisher, and Wright also considered multiple-factor systems. The other was that the environment was assumed to be uniform and unchanging, another gross oversimplification, although variable selection coefficients were sometimes invoked. Because of the first, it was possible to deride "beanbag genetics." Because of the second, this approach seemed to be a formal exercise in math with little relation to the real world.

Further Developments

It is of interest in this connection that the field of population ecology was developing during much the same time period as population genetics, but there was virtually no cross-fertilization between the two emerging disciplines. The population ecologists were much concerned with the effects of the environment on population dynamics (that is, on changes in the numbers of organisms), but whereas the population geneticists assumed a constant environment to simplify their calculations, the population ecologists assumed that all individuals in their populations were identical (black boxes, if you will) so that they could ignore the possible effects of individual variation on such factors as birth and death rates and thus simplify their calculations. Some of the notable figures in the development of population ecology were Joseph Grinnell (1914), Raymond Pearl (1925), A. J. Lotka (1925), V. Volterra (1926), Charles Elton (1927), G. F. Gause (1934), and more recently, G. Evelyn Hutchinson (1957, 1978), but one can search in vain through their writings trying to find signs of any significant influence from population genetics. Similarly, there is little evidence that population geneticists took much notice of population ecology in formulating their theories.

The only workers who attempted to deal with both ecological variation and genetic variation were the ecological geneticists, primarily botanists, such as G. Turesson (1922b), W. B. Turrill (1940), and J. Clausen, D. D. Keck, and W. M. Hiesey (1940), as well as E. B. Ford, who wrote the first book on ecological genetics (1964).

Two other geneticists worth mentioning in connection with the development of modern evolutionary theory are Richard Goldschmidt and H. J. Muller. Goldschmidt's research on geographical variation in gypsy moths and on developmental genetics was widely accepted, but his theories about separate evolutionary mechanisms for microevolution and macroevolution were not taken very seriously, although they continue to be revived periodically (Gould, 1982a). Muller somehow tends to be overlooked in the discussions about the modern synthesis. Although his name will always be associated with mutation, in his writings about evolution he was a strong selectionist. An example of this attitude, and of his tendency to deal with a subject comprehensively and exhaustively, is to be found in his Harvey lecture "Evidence of the Precision of Genetic Adaptation" (1948). Muller is so well known for his work on induced mutations that his wide-ranging other interests are sometimes overlooked, but he wrote on such evolutionary topics as why there are fewer polyploid animals than plants (1925), the origin of reproductive isolation (1940, 1942), evolutionary rates in asexual versus sexual populations (1932a), and genetic loads (1950). Perhaps some indication of the respect he commanded from evolutionists is that he was invited to give the summation for the famous Princeton symposium *Genetics, Paleontology, and Evolution* (Jepsen, Mayr, and Simpson, 1949). Although his writings were sometimes exhausting as well as exhaustive, they were always worth reading.

Following the publication of the seminal interpretive works by Dobzhansky, Huxley, Mayr, Simpson, Darlington, and Stebbins, a rather remarkable meeting of minds on the basic features of the evolutionary process occurred among scientists in a number of disparate fields. There was widespread agreement among systematists, geneticists, paleontologists, botanists, ecologists, and other biologists about the mechanism of evolution. Stephen Jay Gould (1983) called this "The Hardening of the Modern Synthesis," for he felt that all biologists fell into a lockstep, accepting a limited and limiting version of the modern synthesis that actually hampered further development of evolutionary thought. However, Gould looked upon Mayr's 1963 book as the authoritative statement of the modern synthesis even though, according to Haldane and Wright, it was not an adequate account of the synthesis. Like others, Gould also accepted Dobzhansky's 1937 book as the foundation of the modern synthesis, despite Dobzhansky's disclaimers. Even more interesting is Gould's admission that he and Niles Eldredge were ignorant of Wright's work (Gould, 1982b) when they proposed their theory of punctuated equilibrium in 1972. This startling admission helps explain why the theory was presented with such fanfare as a breakthrough in evolutionary thought when in fact it could be easily accommodated within the existing theoretical framework.

In the last thirty years, a number of questions have attracted the attention of evolutionists. V. C. Wynne-Edwards (1962), by urging the case for group selection, raised the more general topic of levels of selection or units of selection. W. D. Hamilton (1964) introduced the concept of inclusive fitness in his efforts to deal with the evolution of social behavior. John Maynard Smith and G. R. Price (1973) used game theory as a new way to approach the study of social behavior. Another subject that has drawn considerable attention is the evolution of sex and the related topics of the cost of sex and the cost of meiosis. Robert Sokal and P. H. A. Sneath (1963) developed the methodology of numerical taxonomy, which provided a new approach to the study of systematics.

The discovery of high levels of biochemical polymorphism in natural populations by R. C. Lewontin and J. L. Hubby (1966) had several ramifications. Motoo Kimura (1968) proposed the neutral theory of molecular evolution to account for the high levels of biochemical polymorphism and the high rate of molecular evolution. The apparent constancy of the molecular evolutionary rate led to the postulation of a molecular evolutionary clock, a matter that, like neutralism, has aroused considerable controversy. The discovery of the existence of introns and exons, of regulatory genes, "parasitic" genes, and "selfish" genes, and of transposable elements, pseudogenes, and repetitive DNA has forced evolutionists to rethink the ways in which molecular evolution can occur. Many of the papers written on these subjects, like Eldredge and Gould's writings on punctuated equilibrium, show a surprising lack of knowledge of earlier evolutionary thinking related to their work. Edward O. Wilson's massive study, *Sociobiology* (1975), similarly seems limited when dealing with basic genetic and population genetic concepts. The thrust of Wilson's book is its emphasis on the importance of hereditary influences in the development of social behavior, but the arguments are weakened by a simplistic presentation of genetic concepts.

By now, it should be clear that my perception of the events in the development of evolutionary thought is somewhat different from the version that has been widely publicized and is becoming increasingly accepted. In recent years there has been a growing tendency to resort to verbal arguments, mathematical models, and computer simulations to settle questions about evolution rather than to test the theories in actual populations. The simplifying assumptions that creep into such work make the results preliminary and suggestive rather than definitive. More reliable, but often more complicated, data emerge from the study of actual populations. The most informative approach to the study of the mechanism of evolution involves studies of living populations in the field and in the laboratory (Nevo, 1978, 1988, 1991; Endler, 1986). Difficult though this approach may sometimes be, it still provides the best answers to questions about how evolution occurs. In the following chapters, we will deal with some of the evolutionary concepts undergirding the modern synthesis and try to develop operational, working definitions of these concepts that can be used in the study of actual populations.

Chapter 2
The Factors of Evolution

In the preceding chapter, many terms were used but not defined because they are familiar enough to pose no great difficulty when used in a historical context. In this chapter, in which the mechanism of evolution is outlined, these terms will be more carefully defined because they often mean different things to different people. Some of these concepts seem to have inherent ambiguities, but the definitions offered here will be working or operational definitions useful to a biologist trying to study evolution.

Darwin proposed that the similarities observed among different species are owing to their descent, with modification, from a common ancestral species, that the resemblances reflect a genetic relationship between species, and that the more similar they are, the closer the relationship and the more recently they have diverged from one another. The basic question for evolutionists, then, is how does a population with one set of characteristics arise from an ancestral population with a somewhat different set of characteristics? Darwin's original definition of evolution was descent with modification. The rise of genetics and then of population genetics made possible a more precise definition of evolution as a change in the kinds or frequencies of genes in populations.

Implicit in this definition of evolution is the question, what factors can produce changes in the kinds or frequencies of genes in populations? The answer is that just four factors can lead to evolutionary change: mutation, selection, gene flow, and random genetic drift. (For a more detailed exposition of the mechanism of evolution see Merrell, 1981.)

Mutation

The ultimate source of genetic variation in populations is mutation, which may be defined as any heritable change in the hereditary material not due to segregation or to genetic recombination. Under this definition, mutations may be genic, chromosomal, or even extrachromosomal. Even though genic mutation rates (the frequency with which a particular mutation occurs per locus per generation) may be very low, mutation pressure (the continued recurrent production of an allele by mutation) can eventually, given sufficient time, produce significant changes in gene frequencies. Thus mutation must be regarded as an independent evolutionary factor because it can produce changes in gene frequency in the absence of selection, gene flow, and random drift.

Selection

Natural selection is a second factor that can bring about gene frequency changes in populations, and thus bring about evolutionary change. We have already dealt with Darwinian natural selection in Chapter 1. The essence of natural selection is that some individuals leave more progeny in proportion to their numbers than others in the same population. Although many deaths and failures to reproduce may simply be the result of chance events, nevertheless the individuals who survive and reproduce are apt to include some who are better adapted than others to survive and reproduce under the existing physical and biological conditions. Thus, selection is phenotypic; the individuals with the phenotypes best adapted to the existing conditions are more likely to survive and reproduce. The critical word here is "reproduce." Although Darwinian natural selection was often referred to as "the survival of the fittest," and even though survival to sexual maturity is necessary for an individual to reproduce, survival alone is not enough for natural selection to occur. The crux of natural selection is differential reproduction.

If gene frequency change and evolution are to occur, however, the phenotypic differences on which natural selection acts must to some extent be hereditary. Only if the individuals with different phenotypes also have different genotypes will the genes and genotypes of the individuals favored by natural selection tend to increase in relative frequency. It is worth stressing that even though we may speak of selection favoring a particular gene or genotype, this effect is indirect, for selection is, in fact, always phenotypic. However, selection can only produce evolutionary change when it acts on a genetically variable population.

Gene Flow

The third independent factor leading to gene frequency change is known as migration or gene flow. In almost every species, the number of individuals is so great that the species population is widely dispersed and subdivided into a number of breeding populations or demes. If each deme were completely isolated from all others, it would constitute a separate evolutionary unit. Ordinarily, however, there is some dispersion and exchange of individuals between demes of the same species, the extent of which is dependent on the population structure of the species—that is, the way the total species population is subdivided in space into separate breeding populations. If the immigrants become established as reproductive members of the population they have joined, and if they differ genetically from that population in the types or frequencies of the genes they carry, then the gene pool of that population will have been changed by gene flow. The influx of a large number of genetically divergent individuals into a breeding population can produce a dramatic, rapid change in the genetic characteristics of that population. Gene flow is an independent evolutionary factor because it can lead to gene frequency change even in the absence of mutation, selection, or drift.

Genetic Drift

The fourth and final evolutionary factor is random genetic drift. Gene flow is dependent on population structure, the way the species population is subdivided in space. Random genetic drift is dependent on population size, or to be more specific, on the "effective" size of the breeding population in individual demes or breeding populations. The smaller the breeding population, the greater the potential for random fluctuations in gene frequency owing to accidents of sampling. We have already noted that the Castle-Hardy-Weinberg equilibrium holds in large populations; that is, gene frequencies tend to remain constant from generation to generation. This is so because the pool of gametes giving rise to a large population is large, and the sampling error or the probability of a sizable deviation from the gene frequencies of the previous generation is correspondingly small. Few gametes are needed to form a small population, however, and the probability of large sampling errors is much greater. The principle involved is very similar to the rules of probability for coin tossing. If you toss a coin 1,000 times, the expected deviation from a 50:50 ratio of heads to tails is very small, of the order of 1 or 2 percent, at most. If, however, you tossed a coin only 10 times, you would not be surprised, I hope, if you got ratios of heads to tails quite far removed from 5 heads and 5 tails. The exact probabilities are only 252 chances in 1,024 that you would get exactly 5 heads and 5 tails, 420 in 1,024 that you would get 4 heads and 6 tails or 6 heads and 4 tails, 240 in 1,024 that you would get 3 heads and 7 tails or 7 heads and 3 tails, and so on down to a probability of only 1 in 1,024 of tossing 10 heads in a row. If these were alleles of equal frequency being drawn from a pool of gametes, there is less than 1 chance in 4 that they will still be equal in frequency in the next generation. This point is being belabored because it makes clear the random nature and unpredictability of genetic drift.

There is a difference, however, between random genetic drift and coin tossing. With the toss of a coin, the probability of heads and tails always remains constant at 50:50. In a population, gene frequencies may range anywhere from 0 to 1, and the existing frequencies form the pool from which the gametes are drawn to form the next generation. One of the consequences of this difference is that when gene frequencies deviate from equality, the less frequent allele always has a slightly greater probability of decreasing still further in frequency than of increasing in frequency. (For proof, see Merrell, 1981, p. 230.) Thus the less frequent allele will always be tending to drift to extinction and the more common allele to fixation. Mayr (1963, p. 213) considered the word "drift" inappropriate for the phenomenon of random genetic drift: "To apply the term 'drift' to nondirectional random fluctuations is unfortunate, since in daily language we generally use the term 'drift' for passive movements in a more or less unidirectional manner." But because the rarer allele does, in fact, always tend to drift toward extinction in a more or less unidirectional passive manner, Wright's use of the word "drift" for the phenomenon fits Mayr's prescription exactly.

Directional, Random, and Unique Events

Evolution, then, results from the combined effects of four factors, namely, mutation, selection, migration, and drift. The first three are sometimes referred to as directional forces, for if mutation rates, migration rates, and selection coefficients remain constant, it is possible, in theory, to calculate the magnitude and direction of the changes in gene frequency resulting from the combined effects of these factors, which are then referred to as mutation, migration, and selection pressures.

In addition to the directional or deterministic aspects of evolutionary change just noted, there is also a random aspect. In part, this is the result of random genetic drift, but other factors may be operative as well. If mutations are recurrent, there seems to be some orderliness to the process in that mutation rates can be determined. Mutations are rare events under most circumstances, however, with an average rate of about one new mutation per million genes per generation. Nor is it possible to predict where or when a spontaneous mutation will occur. Even though mutagenic agents may raise mutation rates, most of them have general effects rather than specific mutagenic effects on particular loci. Furthermore, mutations do not ordinarily occur as adaptive responses to environmental insults affecting the organism. They are random with respect to the environment in which they are occurring. Thus there is a large element of chance involved in mutation as well as in drift. Mutations are rare and unpredictable events. It is also unlikely that selection coefficients or migration coefficients remain constant indefinitely so that fluctuations in the effects of these factors should also be expected.

Finally, some events significant to evolution such as a favorable new mutation, a novel genetic recombination, a hybridization, a selective event, or a population bottleneck may be unique.

The course of evolution then may be affected by a complex assortment of deterministic and chance and even unique events. Because of this complexity, it is rather surprising to find the selectionists constructing such a variety of plausible stories about the adaptive significance of the traits they observe while the neutralists equally plausibly argue that the observed variation is adaptively neutral. As long as the evidence remains equivocal, the positions of the neutralists and the selectionists resemble articles of faith more than well-established scientific facts. The great need in the study of evolution continues to be well-designed empirical studies that permit choices to be made among competing theories.

Other Factors?

Because factors other than mutation, selection, migration, and drift have sometimes been included as part of the mechanism of evolution, it seems advisable to discuss the nature of some of these factors and why they do not belong with the major factors of evolution. Darwin, for example, in *The Descent of Man and Selection in Relation to Sex* (1871), postulated that sexual selection was necessary to explain the evolution of the differences between the sexes and was distinct from natural selection. This idea has persisted (e.g., Campbell, 1972) even though sexual selection is not actually an independent evolutionary force but is, along with viability, developmental rate, longevity, and so on, one of the components of fitness that together determine reproductive success.

Wilson and W. H. Bossert (1971) added meiotic drive as a fifth factor to mutation, selection, migration, and drift. Although the mechanisms for meiotic drive remain poorly known in most cases, it results in the unequal production of different types of gametes, resembling gametic selection in its effects, and thus can be included under the rubric of natural selection.

Stebbins (1971) stated that the modern synthesis recognizes five basic types of process: gene mutation, changes in chromosome structure and number, genetic recombination, natural selection, and reproductive isolation. This list is somewhat different from the factors cited earlier. Genetic recombination and changes in chromosome structure and number result, however, in a reordering of existing genetic material but do not bring about evolutionary change of and by themselves. They may provide new and different combinations of genes on which natural selection may act, but they do not belong in the same category with the factors that can independently lead to evolutionary change. The same may be said of "reproductive isolation." Reproductive isolation may help to ensure the discreteness of breeding populations, but it ordinarily arises as the result of the effects of some combination of mutation, selection, and drift, and should therefore be considered as one of the "accessory processes" mentioned by Stebbins. It is not one of the primary evolutionary factors. Interestingly enough, Stebbins listed migration, hybridization, and chance (drift) as his three accessory processes. In reality, drift and migration (gene flow) are two of the four basic processes that can independently cause evolutionary change. Hybridization, on the other hand, falls within the rubric of gene flow, albeit in its most extreme form, and should not be considered as a separate factor at all.

The mechanism of evolution reviewed here has been just a barebones outline of the modern synthetic theory of evolution. Because it includes both random and directional factors, it cannot be regarded as falling into either the selectionist or the neutralist camp. There is provision within the framework of the theory for both random and nonrandom events, but the actual role and relative importance of directional and chance factors and of adaptive and nonadaptive evolutionary changes must be determined by observation and experiment rather than by theoretical reasoning. In the following chapters we shall consider some aspects of the modern synthetic theory of evolution in more detail.

Chapter 3
The Study of Evolution

To study the evolution of a group we need to study the autecology of that group. The word autecology is used in two senses: to physiological ecologists it means the ecology of a single organism; to population ecologists it means the ecology of a single species. The study of evolution in a group requires knowledge of its autecology in both senses of the word. In particular, we need to stress the characteristics most useful to the study of evolution and how to measure these entities. The discussion will deal with sexually reproducing populations of higher organisms with separate sexes.

Individual Traits

Let us first consider traits of individuals of particular interest to an evolutionist. The first is the individual's genotype. Every individual, except for identical twins, has a unique genotype. At present, it is not possible to identify more than a very small fraction of the genome of any individual. The "struggle to measure variation," as Lewontin (1974) has called it, has been a long and frustrating one, and is far from over. We shall return later to a consideration of what we need to know about the genotype and how such knowledge can be obtained.

The next trait is the individual's phenotype. As with the genotype, searching for the phenotype opens such a Pandora's box of possibilities that we shall defer discussion of the phenotype until later in this chapter.

Other traits of interest include the developmental rate, which determines the time it takes an organism to reach sexual maturity. Related to this is viability, usually measured as the time of death of a developing individual. Viabilities are often reported as the proportion of a group of individuals surviving to a given age, but this figure is then a characteristic of a group rather than of an individual. Longevity is the length of life of an individual, but longevities too may be reported as averages for different groups, and thus become group traits.

Fecundity and fertility are often used interchangeably in discussions of reproductive capacity, but fecundity is best used to refer to the number of eggs a female produces. It is thus a characteristic of an individual female. Fertility can be defined as the proportion of fertilized eggs that produce viable embryos; it thus is somewhat more complex than fecundity because it is a trait dependent on the characteristics of both the male and the female parent.

Mating success can also be measured for individual males and females, although again these measurements are often used to estimate the relative mating success of one group of males or females as compared to another group of males or females. Again, because the mating success of a given individual may depend in part on the characteristics of its mating partner, this is not always a simple trait to measure. These various traits—developmental rate, viability, longevity, fecundity, fertility, and mating success—are all components of fitness, but none alone can be considered to be equivalent to fitness.

Fitness, as we have already stressed, is best measured in terms of reproductive success. All the components of fitness listed earlier contribute to reproductive success, but none alone is a sufficient measure of fitness. The struggle to measure fitness is even more difficult than the struggle to measure variation. J. M. Thoday (1953) has even suggested that a true measure of fitness would require that we measure the persistence of evolutionary lineages over long periods of time—say, 10^8 or 100,000,000 years. Most of us do not have that much time available, however, and must seek some more realistic measure. An operational measure would be to determine the number of sexually mature progeny that an individual contributes to the next generation.

Even such a measurement is not without its practical problems. First, in species with long generation times, it may take years to gather such information. Furthermore, in natural populations it is often difficult to determine with certainty the parentage of individuals in the population, especially the identity of the male parent. Unless the parent-offspring relationships among the members of the population are known, however, it will be difficult to determine how many sexually mature offspring each individual is contributing to the next generation. Such information is difficult to gather even in experimental laboratory populations unless each member can be individually identified and its parents are known. It is little wonder that most estimates of fitness are indirect.

Although we might wish to measure fitness in a dimensionless system, at one time and place, we have just seen that some time must elapse so that it can be determined which progeny reach sexual maturity. Furthermore, because fitness varies with the environment, it must be assessed in relation to some set of environmental conditions, and some way to characterize the environment must be found. Thus the measurement of fitness does not occur in a dimensionless system but must be done over a period of time under known conditions.

The distinction is sometimes drawn between relative and absolute fitness. To measure relative fitness, one phenotype or genotype is taken as the standard against which the fitness of other phenotypes or genotypes is compared. For mathematical simplicity, the reference type chosen is usually the one with the highest fitness. Relative fitness is easier to study than absolute fitness because little effort need be spent determining the environmental conditions. As long as the environmental conditions are the same for the phenotypes or genotypes being compared, it is often assumed that the relative fitnesses remain the same even when the environmental conditions change. For example, when a heterozygote is fitter than its two corresponding homozygotes under one set of conditions, it is often assumed to be superior to the homozygotes under other sets of conditions.

Absolute fitness is studied less often, probably because absolute fitnesses are more difficult to determine. The absolute fitness is usually measured by the number of individuals in a population. A population whose numbers are increasing is taken to have greater absolute fitness than a population whose numbers remain constant or are declining. Such an assumption is only meaningful, however, if the environmental conditions remain constant. Thus even absolute fitnesses are relative, relative to some set of environmental conditions.

Moreover, equating absolute fitness with population size is questionable, as can best be illustrated by an example. If a population of 10,000 flies is exposed for the first time to DDT and all but 10 genetically resistant individuals die, the size of the population has declined precipitously, but its fitness in its new environment has greatly increased. Perhaps it is not surprising that relative fitnesses are reported much more frequently than absolute fitnesses.

Another characteristic of an individual is its ecological niche. In simple terms, to determine the ecological niche of an individual, it would be necessary to determine the limits of the physical environment within which it can survive, all its requirements in terms of food and habitat, and all its relationships to other organisms of its own and other species—a very large order, indeed. This concept of the fundamental niche of an individual can also be applied to a deme or for that matter to an entire species, and we shall have occasion to explore it further.

An additional trait of individuals is their vagility or ability to disperse. To study dispersal, one must be able to identify individuals. If one wishes to study gene flow, for example, some way must be found to detect the movement of individuals between populations and determine whether they have in fact contributed to the gene pool of the population they have joined.

Thus far, we have dealt with some phenotypic traits of individuals, the components of fitness, that may be of particular interest to students of evolution. We have also stressed that natural selection is phenotypic, that the differential reproductive success of individuals with different phenotypes (and genotypes) is the essence of natural selection. At this point, it seems desirable to discuss the nature of the phenotype in order to reach a mutual understanding of what is embodied in this concept.

The Phenotype

In pursuit of their research, all biologists work with organisms of one kind or another. More specifically, they work with the phenotypes of these organisms, but the particular interests of the observer tend to shape his or her perception of the phenotype. A geneticist, a biochemist, a physiologist, a cell biologist, an anatomist, and an ecologist will perceive the same organism in very different ways, for most biologists study only limited aspects of the phenotype. About the only biologists forced to think about the phenotype in broad terms in their work are those dealing with either the process or the products of evolution, the evolutionary biologists and the systematists. But even these biologists, working on such problems as adaptation or classification, sometimes study only a limited number of aspects of the phenotype.

The phenotype of an organism is often equated with its adult morphology, but this view is too narrow, for the phenotype is the sum total of all the traits of an organism. Phenotypic traits range all the way from the genetic system through molecular, cellular, physiological, developmental, and morphological traits to behavior. Thus, both the DNA and the way it is organized into chromosomes as well as the behavior patterns for nest building in birds or web spinning in spiders are phenotypic traits. Even the nest and the web are phenotypic expressions of the genotypes of the bird and the spider. Furthermore, the phenotype of an individual is not static but dynamic, for it changes continuously from fertilization through infancy and adolescence to maturity, senescence, and death. These concepts, long familiar to geneticists, were discussed by Dawkins (1982) in his book *The Extended Phenotype*.

Although a given phenotype is often equated with a particular genotype, this is overly simplistic. The genotype sets limits on the kinds of phenotypes that can develop, but one genotype can give rise to many phenotypes. The particular phenotype that develops is determined by the particular set of environmental conditions encountered by the developing embryo. The phenotypic modifications that occur in a developing organism are generally adaptive responses that enable the organism to cope better with its immediate environment. For example, tadpoles reared in ponds deficient in oxygen will have larger gills than controls reared in ponds with a more abundant oxygen supply, and are thus physiologically better adapted to obtain oxygen in their low O_2 environment.

Single traits are ordinarily influenced by many genes, and because of pleiotropy, one gene may affect many traits. The phenotype that emerges results from the actions and interactions of the genes, the effects of environmental factors, and the interactions between the genotype and the environment. Although the genotype permits a range of possible phenotypic reactions, there are limits to the possible kinds of phenotypes that can develop. A mouse genotype, for example, always gives rise to a mouse, never to an elephant.

Although the influence of the environment on the developing organism is widely appreciated, Lewontin (1982) has recently emphasized the influence of the organism on its environment. He stressed the "interpenetration" of organism and environment as if they were equally important and inseparable in the development of the individual. However, even though an environment can exist independent of organisms, whereas an organism always inhabits an environment of some sort, it seems unwarranted to weight environment and genotype equally and to ignore the primacy of the genotype in the development of an organism. Without a genotype there can be no organism, for the genotype determines, within limits, the kind of organism that develops, something the environment can never do.

The phenotype of an organism consists of an infinite number of observable or measurable traits. Our conception of the phenotype of an individual then depends on the particular set of traits we choose to observe or to measure. External morphology tends to dominate our perception of individuals simply because it is so obvious, but it provides only a limited perspective on the full range of phenotypic traits. We distinguish one individual from another by the differences between them. These differences may be qualitative or quantitative in nature.

Qualitative or discontinuous variations fall easily into discrete classes: red or white flowers, A- or B-type blood, simple or compound leaves. In other words, the expression of the trait is clear-cut, and each individual can be unequivocally classified in one category or another.

Quantitative traits such as height or weight are continuously variable; that is, they are metric traits that can be of any value over a given interval, say, from 9 to 10. Even though an infinite number of values is possible over this interval, limits on the precision of measurement make even these measurements discrete, for example, to the nearest millimeter or milligram. Continuously varying traits tend to have a normal frequency distribution in the form of a bell-shaped curve.

Another type of biological variation is meristic variation in such traits as eggs per clutch, tomatoes per plant, bristles per fly, or scales in the lateral line of fish. In these cases the variation is discontinuous, for it involves only whole numbers or integers, but these traits are usually treated as quantitative traits, both statistically and genetically, because they tend to have a normal frequency distribution like other quantitative traits.

Some traits, such as the rumpless condition in roosters or pneumococcus resistance in mice or schizophrenia in humans, appear to be qualitative and discontinuous, for the rooster is either rumpless or normal, the mouse either dead or alive, and the person either psychotic or normal. Traits like these, however, seem to have a developmental continuity on which a threshold imposes a discontinuity. The underlying continuous variation for these traits is both genetic and environmental in origin like that of other quantitative traits, but the observed trait is discontinuous. The threshold marks the point of discontinuity on an underlying continuous scale of variation. The same reasoning has been applied to meristic variation where it is assumed that the physiological processes involved in scale or bristle formation vary continuously, but that some threshold must be exceeded before an additional scale or bristle can be formed.

Although the distinction between qualitative and quantitative traits is useful, by now it should be obvious that the distinction is far from clear-cut. Quantitative, meristic, and threshold traits all are sensitive to environmental influences and are usually influenced by genes at many loci as well. Qualitative traits are usually more simply inherited, but even for these traits environmental effects may be important. Harelip in mice, for example, is inherited as a recessive with low penetrance; that is, not all the individuals homozygous for the harelip gene show the harelip trait. Penetrance is measured as the frequency in percent with which a gene or gene combination is expressed phenotypically among individuals all having the same genotype. Furthermore, the severity of the harelip-cleft palate syndrome can vary widely; that is, the expressivity or degree of phenotypic expression of the genotype among the affected mice may range from slightly to severely deformed. These variations in penetrance and expressivity may result from both environmental effects and differences in the residual genotype, or, if the residual genotype is held constant, from environmental influences alone.

Extreme environmental conditions may induce phenocopies, that is, environmentally induced, nonhereditary phenotypic imitations of the effects of mutant genes. Despite the abnormal appearance of phenocopies, their offspring will have normal phenotypes when grown under normal conditions, for no genetic change is involved in the production of phenocopies.

Organisms may be characterized by a broad spectrum of traits, including DNA sequences, proteins, karyotypes, metabolic and physiological differences, morphology, and behavior. The base sequence of its DNA might be regarded as the ultimate information about an organism, but in our present state of ignorance, this information tells us very little about how an animal might behave. Conversely, studies of its morphology or behavior do not permit strong inferences about its DNA content.

Given the wide range of possibilities, we may ask which phenotypic traits are most useful in characterizing organisms. In the past biologists have relied almost exclusively on morphological traits. Numerical taxonomists have claimed that the most appropriate way to proceed is to study a large number of characters, randomly selected in order to ensure an unbiased sample of the phenotype of the organisms they are studying. They feel that all traits should be weighted equally, that none should be regarded as more significant than any other. This assumption suffers from the handicap that, as products of differentiation in the developing embryo, some of these traits may be correlated and not truly independent. They then measure one hundred or more morphological traits in each individual in a sample of organisms, subject the data to statistical analysis, and draw their conclusions. This laborious approach has a certain gratifying precision about it, but its biological significance is questionable, for there is a certain mindlessness in such reliance on a single type of trait to reach taxonomic and phylogenetic decisions. More recently, DNA sequencing and electrophoretic studies of proteins have been used to construct cladograms from which relationships could be inferred, but again the conclusions were based on a single type of trait.

Taxonomy remains the cornerstone of all biological research, for if a scientist cannot identify the organisms used in his research, others will be unable to repeat or validate or extend that research because of their uncertainty about the species that was used. More than a century after Darwin, the informed judgment of a competent taxonomist remains our most reliable means of identifying the species with which we work. In laying the taxonomic foundations on which the rest of us build, the taxonomist would do well to obtain information about every possible aspect of the phenotype, in its broadest sense.

The working taxonomist can draw on information from many kinds of phenotypic traits to make decisions about phylogenetic relationships among the organisms under study. Most taxonomic work is still based primarily on morphological traits, however, and more particularly on the external morphology of adult forms, although internal morphology is sometimes used. In some cases comparative embryology is more helpful than comparative anatomy because affinities are clearer in the early stages of development than in the adults, but the basis of comparison is still morphological.

Recently, phylogenies based on other types of data have become popular. Changes in the number, size, shape, and banding structure of chromosomes have been used to determine phylogenetic relationships, especially in *Drosophila* where overlapping inversions have been particularly useful. Most of the other types of data come from studies of macromolecules such as proteins and nucleic acids. Immunological studies of proteins date from the early days of the twentieth century, when G. H. F. Nuttall (1904) used this approach to tackle such puzzling taxonomic problems as the relationship of whales to other mammals and of horseshoe crabs to other arthropods.

Radioimmunoassay has brought this technique to a new level of sophistication (Lowenstein, 1985). The most widely used method for the comparison of proteins in different species is undoubtedly gel electrophoresis because it is relatively simple and inexpensive. Amino acid sequencing of proteins gives more detailed information about their structure but is less widely used because it is time-consuming and costly. Studies of nucleic acids may involve DNA hybridization, DNA or RNA sequencing, or more recently, studies of mitochondrial DNA. The information derived from the work on macromolecules gives a different and perhaps independent perspective on the relationships among species from that provided by morphological comparisons.

Evolutionary biologists have recently been preoccupied with the adaptive significance (or lack thereof) of traits in the organisms they study. The study of adaptation lacks the long tradition that governs the study of taxonomy, and suffers from too much armchair speculation and too little effort to actually measure the relative fitness of different phenotypes. Like the taxonomists, the students of adaptation seem to have a limited perspective on the phenotype. The weaknesses of present studies are often traceable to this narrow view of the relation between the phenotype and fitness where attention is focused, for example, on a single morphological difference or a single segregating enzyme locus. Such studies are necessary first steps, for the experimental study of natural selection on the phenotypes of individuals in wild populations is still in its infancy in most respects. However, the realization that it is possible to study, rather than speculate about, the effects of natural selection on phenotypes has ushered in a new era in the study of evolution, for phenotypic selection results in evolutionary change when different phenotypes involve different genotypes, and a clear understanding of the nature of the phenotype and its relation to the genotype is essential for further progress in the study of adaptation and evolution.

Group Traits

Thus far, we have discussed the nature of the phenotype of individuals because the basic unit on which selection acts is the individual and its phenotype. The basic unit of evolution is the breeding population or deme, however, and now we need to consider what kinds of parameters we must measure to study the evolution of populations. The species is sometimes defined as the most inclusive breeding population of a group of organisms, and our discussion of these parameters can be regarded as applying to a deme or even to a species.

One basic fact about a population is its distribution, both its geographical distribution and its microgeographical or topographical distribution. Knowledge about distribution, both coarse and fine-grained, will help to identify the physical and biological factors that determine where the population is able to live, that is, to learn something about its ecological niche. A comparison of the physical and biological factors at the center of the species range, in marginal populations, and just beyond the limits of the species distribution may help to identify the factors critical to the existence of the species. This sort of information is usually easier to determine for plants than animals.

A second basic fact about a population is its abundance, or the number of individuals in the population. This can be measured as the total size of the population (N), a census figure, but sometimes it is more meaningful to measure the population density or the number of individuals per unit area or per unit volume. A complete census is seldom possible for a natural population, so that a capture-mark-release-recapture experiment is often used to estimate population size. Density estimates at various points within the range of the species may also help to identify the most favorable ecological conditions for the species.

Another measure of population size, essential to the study of random genetic drift, is the effective size of the breeding population (N_e). Very few estimates of N_e have ever been made in natural populations because it is so difficult to determine. The reason is that the effective size rarely equals the actual size of the breeding population. Specifically, the effective size is equivalent to the actual size only in a moderate-sized population with equal numbers of males and females in which mating is at random and all individuals contribute genetically to the next generation. Moreover, in this ideal population, the population size must remain constant; that is, the average family size should equal 2, and the distribution of family sizes should fit a Poisson distribution in which the mean equals the variance.

If a breeding population deviates from an "ideal" population because of unequal numbers of males and females or periodic fluctuations in size, for example, the actual size of the breeding population cannot be used to estimate drift. Instead, the actual size must be converted to an abstract number (N_e) equivalent to the number in an "ideal" population that would have the same rate of decrease in heterozygosity or drift as the observed population. Given the difficulties of determining the sex ratio and family size in wild populations and whether mating is at random and each individual contributes genetically to the next generation, it is small wonder that few estimates of effective population size have been made in natural populations. The effective size of a breeding population may be quite different from the actual census figure and is generally smaller.

The population of an entire species seldom forms a single large breeding population but instead is usually subdivided into a number of separate breeding populations. Even if the species population is not subdivided, the range of the species is usually so great that individuals in different parts of the range have no opportunity to mate because they are isolated from one another by distance. A gray squirrel in Missouri, for example, will not mate with any gray squirrels in Maryland because of the great physical distance separating them. Thus, the population structure, or the way the species population is distributed in space, may be of evolutionary significance.

In the discussion of population structure, the word deme will be used to refer to a local breeding population whereas the more general expression, Mendelian population, will be used to refer to a system of individuals united by mating or parentage bonds (i.e., a reproductive community sharing a common gene pool). Thus, a Mendelian population may be as small as a deme or as large as an entire species. *Polymorphic* variation refers to the genetic variation within a single deme.

The term *sympatric* means living in the same country and refers to members of the same deme or to individuals that are at least potential mates because of their proximity to one another. A somewhat different definition is that sympatric organisms are those found within the area bounded by the points where an individual and its progeny were born. This definition provides a crude measure of the area within which the potential mates of the individual must have lived. *Allopatric* means living in another country and refers to individuals living so far apart that they are not potential mates. *Parapatric*, meaning living in adjacent areas, refers to populations that are geographically contiguous so that mating is possible along their zone of contact. Different demes frequently differ genetically from one another, and this interdemic variation is often called *polytypic* variation in contrast to polymorphic variation, which is intrademic.

The terms sympatric, parapatric, and allopatric are usually used to refer to populations. However, if one tries to determine whether populations are sympatric, parapatric, or allopatric, one must deal with the distribution and behavior of individuals, and the situation becomes confused. In particular, parapatric is used to refer to contiguous populations that may hybridize along their common boundary. However, the individuals in these populations that encounter or actually hybridize with individuals from the other population are sympatric by definition; those that do not are allopatric. Therefore, the so-called parapatric populations are composed of sympatric and allopatric individuals, and the word parapatric becomes rather ambiguous.

Population structure is of evolutionary significance because completely isolated demes of the same species can pursue an independent evolutionary course and may diverge from one another as the result of the combined effects of mutation, selection, and drift within each deme. Demes of the same species are seldom completely isolated from one another, however, and it is necessary to determine the amount of gene flow between different demes, something easier said than done. What is needed are measurements of both immigration rates and emigration rates to and from a given deme, which can best be determined from studies of the dispersal of individuals. Such information alone is not a sufficient measure of gene flow, however, because some measure of the genetic contribution of the immigrants to the gene pool is also necessary. This contribution is most easily measured if the immigrants carry distinctive alleles not found in the population they have joined. One reason evolutionists have devoted considerably more attention to the study of the effects of mutation and selection in natural populations than to gene flow and random genetic drift is that, even though mutation rates and selection pressures are difficult to measure in wild populations, the effects of gene flow and drift are even more difficult to assess.

Another group characteristic of a population is its age structure. Just as the geographical distribution of a species population can reveal something about the ecology of the species, the age distribution of individuals in a population can reveal something about its current reproductive status and its potential for future growth. If a population has relatively many young individuals, its numbers will tend to increase, while a population declining in numbers will have relatively more old individuals, and a stable population will have a more even distribution of age classes.

A population living in a constant environment will tend to reach a constant population size, which means that birth rates and death rates will be equal. It will also have a constant age distribution, no matter what the initial age distribution may have been. Furthermore, this age distribution can be determined from life history data if values of l_x (the proportion surviving to age x) and m_x (the age-specific birth rate) are known. Once a population approximates this age distribution, perturbations may occur because of fluctuating birth or death rates, but the population will tend to return to the stable age distribution spontaneously. This concept can be useful because comparisons between the age distributions actually observed in natural populations and the theoretical stable age distribution permit inferences to be drawn about the rate of population growth or decline. It must be added, however, that the environment of natural populations is seldom constant for any lengthy period of time, so that population size tends to fluctuate rather than remain constant, and the idea that there is a fixed number of individuals, K, representing the carrying capacity of a particular environment is of more theoretical than practical interest.

Thus far, in our discussion of populations, we have dealt primarily with the measurable parameters of static populations at a single point in time: their geographical distribution, their abundance, their age structure, and their subdivision into breeding populations. Populations are dynamic rather than static, however, and for the study of population dynamics, further information is needed. In particular, estimates of the birth rate, the death rate, and the rates of immigration and emigration in a population are necessary if an estimate of the rate of growth of the population is to be made. As might be imagined, as a practical matter it is not easy to determine birth and death rates or immigration and emigration rates in wild populations.

Generation length is a crucial figure in many of the calculations involving population dynamics. It can be defined as the time interval from the midpoint in the reproductive span of one generation to the midpoint in the reproductive span of the next generation. Crude estimates of generation length can be based on longevity data alone, but estimates based on those individuals who succeed not just in surviving but in contributing adult progeny to the next generation are obviously much more difficult to determine, especially in species that reproduce not just once in a lifetime, but repeatedly.

An important parameter in population dynamics is the instantaneous rate of natural increase (r), also called the instantaneous coefficient of population growth. Related to it is the intrinsic rate of natural increase (r_{max}), a hypothetical maximum rate of population growth under optimal environmental conditions. The maximum value of r is often referred to as the biotic potential, and the difference between the biotic potential and the actual rate of increase in a population is taken as a measure of the environmental resistance, that is, the effect of all the limiting environmental factors that keep the biotic potential from being realized. Because the determination of r is dependent on measurements of birth and death rates and of immigration and emigration rates in natural populations, which are never easy to make, not many estimates of r are available for wild populations.

If we are going to study evolutionary change in a population, we need to know something about the genetic changes that are occurring in the population. Ideally, we would like to know the complete genotype of each individual in the population and its fitness relative to all the other individuals in the population. Because each individual in a sexually reproducing population (apart from identical twins) will have a distinctive genotype, it should also have a distinctive phenotype and can be expected to differ from the others in fitness. Unfortunately, this ideal cannot yet be realized, and our knowledge of the genotype of individuals in a population is generally limited to one, or at best, a handful of loci. In order to study these loci in natural populations, it is necessary to determine the gene frequencies (i.e., the relative proportions of the different alleles at each locus) and the genotypic frequencies in the populations. In this way it is possible to follow changes in gene and genotype frequency and attempt to draw inferences about the causes of the gene frequency change.

The study of changes in gene and genotype frequencies in populations seems to have led to the concept of levels of selection or units of selection. The idea is that natural selection operates at a number of different levels — gene, molecule, cell organelle, cell, organism, group, and so on. The fact that we monitor the changes occurring in populations by following changes at the genic, molecular, or cellular levels, for example, does not mean that natural selection is operating at these levels. As pointed out previously, natural selection operates on the phenotype of individuals. If a particular molecule or cell type enables an individual to survive and reproduce better than others with different molecules or types of cells and these differences are heritable, then those genes will tend to increase in frequency, but the selection is not genic or molecular or cellular; it is selection among individuals with different phenotypes and genotypes.

For the study of evolution in a population, the group parameters needed are estimates of gene and genotype frequencies, mutation rates, migration coefficients for the estimation of gene flow, estimates of N_e to assess the possibility of genetic drift, and estimates of selection coefficients or fitness. The relation between fitness and the selection coefficient is best defined by a simple equation: $W = 1 - s$, where W stands for fitness and s, the selection coefficient, is a measure of the selective difference between the gene or genotype in question and some reference gene or genotype. (For more details, see Merrell, 1981, pp. 74 ff.) Additional useful information about the population is the genetic system (the mode of organization and reproduction of the genetic material in the population) and the sex ratio (the relative proportions of males and females). Like many of the parameters mentioned earlier, none of these parameters is easy to estimate in wild populations. Nevertheless, these are the things we need to know if we wish to study evolutionary changes in natural populations. To learn this much about a species and its breeding populations is a demanding, but not impossible, job. A study of the autecology of the species in the broadest sense plus a knowledge of the genetics of its breeding populations is essential if we are to understand the forces acting on the species. Only through a close familiarity with the species in all its many dimensions can we hope to understand its evolution.

Chapter 4
Adaptation, Natural Selection, and Fitness

For a long time, biologists seemed to be generally agreed that adaptation was important in evolution. In Sewall Wright's words (1949), "The process of adaptive evolution is one of central interest from both the philosophical and scientific points of view." Simpson (1950) wrote of "adaptation, an evident and universal fact in nature, as the prime problem that *must* be faced in any attempt to explain evolution." Later he stated (1953), "Adaptation is intimately involved in almost all evolutionary processes." Huxley (1954) noted that "all evolution takes place in relation to the environment, including the biological environment, and its changes. There is a universal process of adaptation." Williams (1966), in his book *Adaptation and Natural Selection*, wrote that "evolutionary adaptation is a phenomenon of pervasive importance in biology." Lewontin (1968) stated that "the study of adaptation is the nexus of population ecology, population genetics and development," whereas Waddington (1968) wrote that "the major problem of the whole of evolutionary theory is to account for the adaptation of phenotypes to environments." Given the weight of all this scientific opinion, perhaps it is not surprising that the evolutionary importance of adaptation has periodically come under attack.

Some of the better-known recent attacks have come from Kimura (1983) with his neutral theory of molecular evolution, from Gould and Lewontin (1979), who mounted a full-scale attack on the "adaptationist programme," and from such authors as Stanley (1979), Eldredge and Gould (1972), and Gould (1980), who feel that adaptive evolution plays an insignificant role in the origin of species and that new paradigms must be sought. All is not lost, however, for such authors as J. R. G. Turner (1983) and A. R. Templeton (1982) have come to the defense of adaptive evolution, with Templeton writing, "Consequently,

I will argue that the interment of adaptation is premature and that adaptation is still alive and well and playing a critical role in the origin of species in both the anagenetic and cladogenetic senses."

Adaptation

It may be surprising that a phenomenon so widely accepted as a fundamental part of evolution should come under attack. Adaptation turns out to be a rather elusive concept, however, and the word has several somewhat different though related meanings, which sometimes leads to confusion. Attempts by biologists and by historians and philosophers of science to define the meaning of adaptation in its various guises have often seemed merely to compound the confusion (e.g., Sober, 1984). Some of these difficulties have arisen because of efforts to work out the relationships among adaptation, natural selection, and fitness. Therefore, it seems worthwhile to discuss the various meanings of adaptation so that we will have a mutual understanding of what is meant when the word is used here.

One way to think of adaptation is as a change in structure, function, or behavior that produces a better adjustment by an organism to its physical and biological environment. Under this definition it is organisms that are adapted to their environments, and more specifically, adaptation is a phenotypic characteristic of an organism. Gould and Lewontin (1979) state that under the adaptationist program, an organism is considered to be a conglomeration of traits, each of which is molded separately by natural selection to maximize the fitness of the organism, and that biologists then concoct "Just So" stories to explain the adaptive significance of each trait. This simplistic version of adaptation is a convenient straw man, however, and often Gould and Lewontin merely substitute one type of "Just So" story for another. For example, they state, with respect to the short front legs of *Tyrannosaurus*, "It is likely to be a developmental correlate of allometric fields for relative increase in head and hindlimb size," which is simply a nonadaptive "Just So" story. Gould and Vrba's suggestion (1982) that the pseudogenitalia of female hyenas are related to high levels of circulating androgens rather than an adaptation for the meeting ceremony merely substitutes one unsupported hypothesis for another.

More surprising is Gould and Lewontin's statement (1979) that "a mutation which doubles the fecundity of individuals will sweep through a population rapidly. . . . Natural selection at all times will favour individuals with higher fecundity." This statement, as it stands, is simply untrue. It might be true if all other things were equal, but as Lewontin has written (Levins and Lewontin, 1985, p. 78; see also Lewontin, 1979a), "all other things are never equal." A mutation that doubles the fecundity of individuals will not necessarily sweep through a population rapidly nor will natural selection at all times favor individuals with higher fecundity. If, for example, a female produces five eggs per clutch, and a mutation leads to a female that produces ten eggs per clutch, this mutation will double the fecundity of the female carrying it, but it will not necessarily sweep through the population. Fecundity is a component of fitness but is not fitness itself. If a female laying five eggs raises three offspring to sexual maturity while a female laying ten eggs raises only one of her offspring to sexual maturity, natural selection will not favor the mutation for greater fecundity and larger clutch size but instead will favor the females whose clutch size results in the greatest number of young successfully reared to sexual maturity. The *ceteris paribus* argument is widely used in theoretical biology and in biological modeling, but if all other things are never equal, and I agree that they never are, then such an assumption, whenever encountered, should be greeted with considerable skepticism.

The first point to be made about adaptation is that, despite its critics, the phenomenon is real and universal. All living organisms are adapted to their environments; if they were not, they would die. This statement amounts to a biological truism. The second point is that it is organisms that are adapted, not traits. To think of an organism as a conglomeration of adaptive traits is inappropriate. A particular trait may enhance the ability of an organism to survive and reproduce (i.e., the measure of adaptation), but it is the organism that survives and reproduces, not the trait. The propensity to view the adaptive significance of individual traits separately has opened the door to the critics of adaptation. Just as it is incorrect to equate one gene with one trait when many genes contribute to the manifestation of any trait, it is also incorrect to equate adaptation with a single trait when many traits — in fact, the entire phenotype of the organism — determine how well it is adapted to its physical and biological environment.

Earlier we defined adaptation as a change in the structure, function, or behavior of an organism that produces a better adjustment by the organism to its physical and biological environment. In the previous paragraph we stated that the best measure of how well an organism is "adjusted" to its environment is its ability to survive and reproduce in that environment. Furthermore, all living organisms must be adapted to their environments or they would not be alive, but some may be better adapted than others, as measured by their ability to survive and reproduce. Adaptation is very much a relative concept; the adaptation of an organism is measured not only relative to that of other individuals but also relative to a particular set of environmental conditions.

Adaptedness, Adaptation, and Adaptability

Adaptation in the preceding sense, that of being adapted to a given environment, has been called "adaptedness" by Dobzhansky (1968), and is the product of evolution. He reserved the word "adaptation" for the evolutionary process by which an organism becomes adapted to its environment. He used "adaptability" to refer to the ability of an organism to adapt to changing environmental conditions. These distinctions are useful because "adaptation" is often used to refer to all three phenomena, the process by which organisms become adapted, the products of the adaptive process, or less often, to the ability of an individual to make adaptive responses to its environment.

An example may help to clarify these differences. A polar bear is adapted to survive and reproduce in the cold arctic environment; it is in a state of adaptedness. According to Dobzhansky, adaptation is the evolutionary process that produced the adaptedness of the polar bear.

The concept of adaptability encompasses a variety of individual adaptive responses, some of them rapid responses to environmental stimuli, others much slower. For example, the pupils of our eyes contract in bright light but dilate in darkness. In this case, adaptability leads to visible morphological changes. In other cases, the ability to adapt involves self-regulating processes that maintain constant internal conditions or physiological homeostasis despite external changes. To maintain a constant body temperature, we shiver when cold and sweat when hot. Blood sugar and pH are maintained within very narrow limits despite dietary and other factors that tend to change the glucose and carbon dioxide (CO_2) concentrations of the blood. In these cases, the body's adaptability permits it to maintain a relatively constant temperature and CO_2 and blood sugar levels despite changing environmental conditions. These adaptive responses occur rather rapidly, but other manifestations of adaptability require more time.

Some mammals in the temperate zone, in response to the onset of cold weather, develop a much thicker coat, which they shed when the weather again turns warm. This type of adaptability takes considerably longer than the pupillary response to changing amounts of light.

Although it represents an individual response or adaptation to changing environmental conditions, adaptability, like adaptedness, is a product of evolution and is under genetic control. For example, the golden retriever produces a heavy winter coat, but the Labrador retriever does not because it lacks the genetic capacity to do so. These gradual adaptations to changing seasons are sometimes referred to as acclimatization. All of the examples of adaptability thus far have been reversible, but some cases of developmental adaptation involve an irreversible process.

In many species such as *Drosophila*, development is so canalized or channeled that despite rather drastic environmental insults, the adult organisms all wind up looking much the same. This type of adaptability is known as developmental homeostasis. In other species, especially among plants, there is considerable developmental flexibility, for the same plant genotype will produce quite different phenotypes depending on the environmental growing conditions. Not only do the phenotypes differ, but each phenotype is better adapted to its own particular environment than the other phenotypes possible with that genotype. For example, a dandelion grown at low elevations is a large plant with large leaves and flowers and an erect growth habit, but a clone of the same plant grown in an alpine environment resembles a typical alpine species, becoming a compact dwarf plant growing close to the ground. Thus, developmental flexibility is another type of adaptability. The fact that the developmental flexibility of an organism produces phenotypic changes that are adaptive to the conditions that induce them poses some intriguing developmental and evolutionary questions.

Therefore, adaptability permits an individual to modify its physiology, development, morphology, behavior, or other traits to fine-tune its adjustment to its environment. As the environment is seldom constant, adaptability can be of great importance to the organism. When an individual adapts to changing conditions, no genetic changes occur in the individual, but adaptability itself is under genetic control and must have evolved.

Although fins for swimming or wings for flying are commonly referred to as adaptive traits, it is quite erroneous to think of traits as unitary, separate components of the organism. It is not the trait that is adaptive; it is the organisms with these traits that are adapted to swim or to fly. The question to be asked is not how the adaptive traits evolved, but rather how organisms evolved that were able to swim or to fly by means of fins or wings.

Thus far we have carefully framed the discussion in terms of individual adaptedness, adaptation, and adaptability. However, it is also possible to think of adaptation in relation to populations. Light-skinned human races exposed to sunlight tend to tan, that is, to deposit increased amounts of melanin in their skin as an adaptive protection against sunburn. Dark-skinned human races do not need the stimulus of exposure to sunlight to develop their pigmented skin; they are already adapted in this way. An intriguing evolutionary question is the relation between the adaptability of light-skinned people and the adaptedness of dark-skinned groups. Which is ancestral, and which is derived? How did these differences originate? In any case, it is necessary to think of adaptation in terms of populations as well as individuals because adaptations arise through evolutionary change, and evolution takes place in populations, not in individuals. It is the phenotype of individuals that is well or poorly adapted, but in an evolutionary sense, individuals are ephemeral; continuity resides in the gene pool of the breeding population.

The Evolution of DDT Resistance

At this point, it may be helpful to illustrate some of these concepts by considering some long-term experiments on the evolution of DDT resistance in laboratory populations of *Drosophila melanogaster*. At the time these experiments were begun (Merrell and Underhill, 1956b), some scientists thought that insecticide resistance was a case of adaptability, that the insects became resistant through individual physiological responses to the presence of the insecticide; others believed that the resistance evolved through the action of natural selection on a genetically variable population. These hypotheses were tested by exposing highly inbred laboratory stocks, noninbred laboratory stocks, and stocks derived from wild populations to DDT. The rationale was that if individual adaptability was responsible for the resistance, the genetic composition of the initial populations should not matter, for resistance should appear with equal ease in the homozygous and the more heterozygous populations. On the other hand, if an increase in DDT resistance is owing to natural selection favoring genes for greater resistance in genetically variable populations, then increased resistance should not appear in the inbred homozygous stocks, but only in the more heterozygous laboratory and wild stocks. It was therefore of great interest to find that populations derived from two inbred stocks of *Drosophila*, exposed to DDT over a period of thirty months, failed to become DDT resistant, but that the more heterozygous populations, similarly exposed, developed varying degrees of resistance. Furthermore, flies from the resistant lines reared in the absence of DDT were DDT resistant when tested, but flies from the inbred lines were not. The DDT resistance, then, was not the result of adaptability but rather of the action of natural selection on the gene pool of genetically variable populations to produce flies genetically adapted to their new DDT-laden environment.

The failure of the inbred populations to become resistant during thirty months of selection also suggested that no favorable spontaneous mutations conferring greater resistance occurred during this time, nor did the DDT appear to have induced any mutations for resistance.

Selection for DDT resistance was continued in five populations for more than twenty years. Four populations were each derived from several hundred flies captured from wild populations while the fifth was derived from a well-known noninbred laboratory stock, Oregon-R. An unselected control was maintained for each of the selected populations. The amounts of DDT used at first in the exposed populations were of the order of 0.1 to 0.5 mg on a 3-by-9-cm piece of filter paper, a dosage that killed about 75 to 90 percent of the flies. As the populations slowly became more resistant, this amount was gradually increased to 150 mg. To monitor the level of resistance, separate tests were run in Petri dishes containing DDT but no food to estimate the ED_{50} (ED_{50} = Effective Dose downing 50 percent of exposed flies). These tests showed that the resistant populations had become more than 300 times as resistant as their corresponding controls. Eventually, the resistant lines became so resistant that the testing method had to be changed from probit analysis to a time-based test (Dapkus and Merrell, 1977) because 50 percent of the flies were not dying during the test period and it was no longer possible to estimate an ED_{50}. They were walking around on DDT crystals in the Petri dishes, but were dying of desiccation before they died from the effects of DDT.

A chromosomal analysis (Dapkus and Merrell, 1977) revealed that the resistance in the 91R stock was multifactorial with some factors on each of the three major chromosomes. However, the dominant and recessive effects of genes on the second and third chromosomes were much more important than the effects of genes on the first (X) chromosome, which had no detectable recessive effects. The average dominance of the second chromosome was much less than that of the third chromosome. These large-scale differences in the chromosomes' effects and in their average dominance suggest that a relatively small number of genes for resistance was involved.

Given the origin of this 91R population from a few hundred wild flies from a single locality, this result is not particularly surprising. A relatively small initial population from a single source must have had a somewhat limited amount of genetic variability. Under the rigorous selection pressure imposed by DDT, any genes conferring some degree of DDT resistance were favored and increased in frequency. In such a small initial population, the probability that major genes for resistance might be present was very low. Thus, in this case the resistance increased owing to the gradual accumulation of several factors for resistance, each of relatively minor, though differing, effects. A population confronted by such a drastic selective agent as DDT has no choice; if it does not become resistant, it will die out. Thus, natural selection appears to have seized on whatever genetic variability was available. Continuing genetic recombination, plus, of course, any new mutations for resistance that happen to occur, provides the genetic raw material on which selection acts.

There are several evolutionary lessons to be learned from the evolution of DDT resistance in insects. The first is the opportunism of evolution. A genetically variable population, confronted by an environment containing DDT, may become adapted to these new conditions in any one of a number of ways. The particular route chosen will depend on the genetic variability available on which natural selection can act. Adaptedness can be achieved most rapidly through selection for a favorable dominant for resistance, but if no such genes are available in the gene pool, other genetic routes to resistance will be favored by selection.

A second lesson is the versatility of natural selection. There is not just one phenotype for DDT resistance, but many, for there are many different kinds of phenotypic changes that enable an insect to withstand the effects of DDT. Natural selection, then, has a number of phenotypic options to choose among, and the particular resistant phenotypes that emerge may be quite different in different populations. Hundreds of insect species have evolved resistance to DDT in the decades since it was first used in the 1940s (Crow, 1966; Georghiou, 1969).

A final lesson is the constancy of natural selection. Organisms always exist in some sort of physical and biological environment, which means that they are constantly being tested by natural selection. The other evolutionary factors—mutation, gene flow, and random genetic drift—may or may not be of importance in a given population, but natural selection is always there, a constant presence. This is not to say that the selection pressures themselves are constant, for as the

conditions change, selection pressures will change, but rather that selection of some kind will always be at work in the population.

Natural Selection and Adaptation

We have already defined both adaptation and natural selection, but
now we need to explore the relationship between the two. Natural
selection is phenotypic; in a given physical and biological environment, some individuals are better able to survive and reproduce than
others. If the phenotypic differences that enable them to thrive better
under existing conditions are to any extent the result of genetic differences, the favorable genes will increase in frequency in the population.

Earlier we discussed adaptation as either the process by which organisms adjust to their environments, the product of such adjustments, or the ability of organisms to make such adjustments. All of
these phenomena are under genetic control, and all are the result of
natural selection. In other words, natural selection is the mechanism
for the process by which adaptation occurs; natural selection gives
rise to the products of evolution known as adaptations; and natural
selection is responsible for the adaptability of organisms. Adaptation
in its many guises is manifested in the phenotypes of organisms; selection acts on the phenotypes of organisms; and natural selection is
the evolutionary mechanism that produces adaptation.

Evolutionary changes resulting from mutation, migration, or drift
have been referred to as "fortuitous benefits" (Sober, 1984) if they are
favorable to the organisms under existing conditions, the implication
being that adaptation does not necessarily arise by natural selection.
However, even though new, favorable phenotypic traits may emerge
through mutation, migration, or drift, if they have beneficial effects
on survival and reproduction (the only meaningful meaning of "favorable"), then natural selection is operating, and the concept of "fortuitous benefits" is gratuitous.

As pointed out previously, natural selection is a constant presence
in every population, large or small, isolated or not, regardless of its
mutation rate. This is not to say that nonadaptive evolutionary
changes never occur, but merely that it is just as erroneous to concoct
nonadaptive as adaptive "Just So" stories. Ever since Sewall Wright
first suggested the possibility of random genetic drift, for example,
drift has been a favorite explanation for observed differences between
small, isolated populations, despite a dearth of supporting evidence.

If adaptation is always the product of natural selection, and not other evolutionary factors, it may be asked whether natural selection always produces adaptation. In certain cases of meiotic drive, the driven allele, which is represented disproportionately among the gametes and thus has enhanced reproductive fitness, is invariably associated with reduced survival or viability. The upshot is that with some types of meiotic drive, natural selection would seem to lead, not to adaptation, but to extinction. This apparent dilemma can be dealt with in two ways. One is simply to accept that all adaptations arise as the result of natural selection, but that not all natural selection leads to adaptation. The other is to recall the dual meaning of natural selection, that natural selection favors those organisms best able to survive and reproduce under the existing conditions. Since in these cases of meiotic drive, the organisms best able to reproduce are not the ones best able to survive, it could be argued that these are not really examples of natural selection at all.

Natural Selection and Fitness

In earlier discussions of fitness, it was pointed out that fitness is a phenotypic trait of an organism best defined and measured as the number of its progeny that reach sexual maturity. Fecundity, fertility, viability, developmental rate, longevity, and mating success are all components of fitness and are more easily measured than fitness, but they should not be substituted for fitness itself. Although a distinction is sometimes made between relative and absolute fitness, in truth all fitnesses are relative because even the so-called absolute fitnesses must be determined relative to some environment.

This operational definition of fitness suggests that there is an intimate relationship between natural selection and fitness. Natural selection acts on the phenotypes of organisms to determine which phenotypes are best able to survive and reproduce under the existing conditions. The fittest individuals are not the strongest, the healthiest, the most attractive, or those who live the longest, but, by definition, those who leave the most adult offspring. Rather than a tautology or a hypothesis to be proved, the relationship between natural selection and fitness becomes a matter of definition. Natural selection is the mechanism that increases fitness and enhances the adaptation of individuals to the existing physical and biological conditions. (Recall the relationship between fitness and selection shown by the equation $W = 1 - s$, where W is fitness and s the selection coefficient.) This way of viewing the relationships among natural selection, adaptation, and fitness seems the most serviceable, and it avoids some of the tortured and tortuous arguments sometimes presented on the subject.

We have stressed that natural selection acts on the phenotypes of individuals. A wild population of leopard frogs (*Rana pipiens*) containing many frogs with a unilateral deformity of the hind leg provides a simple illustration of natural selection in action. The deformity ranged from deformed toes through the absence of a foot or lower leg or even the entire leg in some cases (Merrell, 1969a). In late July, the frequency of the deformity among the newly metamorphosed frogs was about 15 percent in a population estimated to number at least 70,000. Despite their deformities, these young frogs seemed otherwise to be vigorous individuals that were as active as their normal cohorts but somewhat less agile; that is, they jumped as often but not as far as the normal frogs. Nevertheless, two months later, in late September, the frequency of deformed individuals had fallen to less than 4 percent. No genetic analysis was made, so the cause of the defect is unknown. If the deformities were environmentally induced, it may be questioned whether this is a case of natural selection at all, or only in a trivial sense. However, one of the most frequent observations about wild populations is their phenotypic uniformity. Despite the wealth of genetic variability known to exist in wild populations, the "wild type" seems to be uncommonly common. The rate of elimination of the deformed frogs suggests why this is so, for in a species that normally requires two years to reach sexual maturity, 73 percent of the deformed frogs had been eliminated in just two months, and it seems a fair assumption that very few if any of the deformed frogs would survive for two years. The cause of their disappearance is unknown though predation is the most probable explanation, for populations of young frogs are subject to heavy predation, especially by the populations of garter snakes (*Thamnophis*) that live in their midst in the summer. In any case, strong phenotypic selection did occur. Whether or not the deformity had a strong genetic basis was immaterial to the frogs; the normal ones survived, the deformed ones did not. The better-adapted individuals would survive to reproduce because of the phenotypic selection that was occurring. Whether this selection would have any lasting effect on the gene pool of the population would depend on the extent to which the deformity was influenced by genetic factors, but the selection pressure was exerted directly on the phenotypes of the individuals. Any selection pressure on the genotypes or genes in the population was indirect, the result of phenotypic selection.

However, if phenotypic selection is to have any long-term evolutionary effect on a population, the phenotypic differences must reflect genotypic differences, and gene frequencies in the population must change. Then the genotypes generated in the gene pool of the next generation will produce an array of phenotypes better adapted to the existing conditions than those of the previous generation. In other words, natural selection is a directional force in evolution, and the direction is toward improved adaptation to the environment. This rather strong adaptationist statement having been made, it must be added that gene frequency change (i.e., evolution) results from the combined effects of mutation, migration, and genetic drift as well as selection, and when gene frequency changes occur in populations, they are not necessarily the result of natural selection nor are they necessarily adaptive. The adaptive significance of observed changes in gene frequency should not be inferred but must be independently established. In the next chapter we shall consider the evolutionary significance of mutation, migration, and genetic drift, the three factors that can produce nonadaptive evolutionary change.

Chapter 5
Nonadaptive Evolution

In our earlier discussion of the mechanism of evolution, we pointed out that four independent factors can lead to evolutionary change: natural selection, mutation, gene flow, and random genetic drift. The first three are often referred to as evolutionary pressures because if selection coefficients, mutation rates, or migration coefficients are known, the direction and magnitude of the expected changes in gene frequency caused by these factors can be calculated. (The essentials for such calculations are given in Merrell, 1981.) Although natural selection is often regarded as the primary cause of directional evolution (e.g., toward improved adaptation), the other two pressures, mutation and gene flow, may also lead to predictable, directional changes in gene frequency. Hence, directional evolution should not automatically be equated with adaptive change resulting from natural selection. However, given the low mutation rates that usually prevail, significant short-term directional changes in gene frequency are unlikely to result solely from mutation pressure. Gene flow, on the other hand, could be quite another matter; a major influx of genes into a population can lead to rapid directional shifts in gene frequency quite unrelated to adaptation.

In this context, it should be noted that most of the calculations in population genetics are based on the assumption of recurrent mutation, recurrent migration, and recurrent or constant selection pressure. Only if recurrent mutations are occurring, so that mutation rates can be estimated, or if recurrent migrations are taking place, so that migration coefficients can be obtained, or if environmental conditions are sufficiently stable for selection coefficients to be relatively constant, are such calculations possible. Perhaps it should be added that genetic recombination, including crossing over, is also a recurrent phenomenon. Only because of the recurrent nature of these phenomena is it possible to construct models and attempt to make predictions about the direction and magnitude of gene frequency changes. In a sense, population genetics has been preoccupied with this aspect of the genetics of populations simply because it is possible to make such calculations and predictions.

One of the attractions of this approach to the genetics of populations is that it reduces the treatment of such biologically disparate factors as mutation (which reflects changes in the DNA molecule), selection (which acts on the phenotypes of individuals), migration (which involves the movement of individuals, or gametes in wind-pollinated plants, from one population to another), and random drift (which is tied to population size) to a single common denominator, namely, their effects on gene frequency change. For example, in discussions of genetic load, mutation and migration are often classified together as the input load, for both are responsible for introducing new genes into the gene pool even though biologically they are quite different. If we confine our attention to a single locus, as is often the case in theoretical calculations, the effects of mutation and gene flow appear to be very similar. However, if an individual carries 50,000 loci mutating at a rate of 1×10^{-6}, only 5 percent of the loci would be expected to mutate, mostly to deleterious mutants, but a migrant who joins a breeding population adds an entire genome to the gene pool and may introduce new and different alleles at a number of different loci in one fell swoop. Thus, in the short run the impact of gene flow on the gene pool of a population may be considerably greater than that of mutation.

Chance and Evolution

Not all evolution is directional, for the course of evolution may be influenced by a number of unpredictable, chance events. This nondirectional form of evolution has been called stochastic evolution, nonadaptive evolution, or non-Darwinian evolution. These expressions reflect certain preconceptions on the part of their authors. Non-Darwinian, for example, implies that Darwinian evolution is equivalent to evolution by natural selection, but Darwin was never that simplistic. Non-Darwinian or nonadaptive evolution is often equated with random genetic drift in effectively small breeding populations, but even with random drift, if the effective size of the breeding population (N_e) is known, it is possible to estimate the expected rate of decrease in heterozygosity in the population. Furthermore, as pointed out earlier, the less frequent allele has a slightly greater probability of being lost than of being fixed so that drift in one direction does tend to occur. Some limits to the range of possible fluctuations in gene frequency can be estimated for a given breeding population size, but expected changes in gene frequency resulting from drift cannot be calculated from generation to generation in the same way they can for selection, mutation, and migration.

Chance can enter the evolutionary picture through a variety of paths, however. For example, instead of having a fixed small population size, a population may fluctuate in numbers sporadically, only occasionally dropping to a size where significant genetic drift is likely to occur. Or a population may go through a severe bottleneck in numbers as a rare or even unique event. These sporadic or unique reductions in population size may have dramatic effects on the variability existing in the gene pool of a population, but may be undetectable and unknown because they occurred in the distant past.

Because mutation rates are so low, of the order of 1×10^{-6}, and because the size of breeding populations is finite and usually much smaller than one million, mutations can probably best be thought of as rare or even unique events in a breeding population rather than as a recurrent pressure. Certainly, in our present state of knowledge, it is not possible to predict with any degree of assurance which mutations are apt to occur or to become established in a given population. Thus, a large element of chance determines the particular array of mutations found in a population, and even if two populations initially are genetically identical, they will tend to diverge from one another in time because different arrays of mutations will become established in the two groups. The genetic divergence between two different sublines of a supposedly homozygous inbred line of mice seems most readily explained on this basis.

Mutations in natural populations are often referred to as spontaneous, which is in one sense a confession of our ignorance as to their causes. In another sense, the term reflects the fact that these mutations are chance events unrelated to the prevailing environmental conditions, and certainly are not adaptive responses to those conditions. As mentioned earlier, induced mutations rarely appear to be adaptive responses to the inducing agent. Thus, even though mutation is considered one of the directional forces in the genetics of populations, as it surely is when mutations are recurrent and build up mutation pressure, nonetheless mutation also has a random aspect and may be responsible for some of the nonadaptive, nondirectional evolutionary changes that are observed. Genetic recombination is another phenomenon that involves an element of chance. Recombination events may be rare because of close linkage or because genes at a number of loci are involved. Then, as with rare mutations, if the recombinant genotypes become established, the course of evolution may be influenced in unexpected and unpredictable ways.

Migration is another factor that may contribute to the stochastic nature of evolutionary change. If, instead of being recurrent, the migration of an individual successful in joining the breeding population is a rare or unique event, it may have a profound impact on the gene pool, but the effect may be incalculable with the usual equations. Even less amenable to estimation are the effects of the introgression resulting from hybridization between members of different subspecies or species. Introgression and hybridization are, of course, merely additional forms of gene flow, but because they do not normally occur, they must be regarded as still another way that stochastic events may influence evolution.

Chance and Natural Selection

Even natural selection, the prime mover in directional evolution, may produce some unexpected twists and turns in the evolutionary path. Selection pressures are not necessarily fixed but may change, sometimes quite abruptly and in quite unexpected ways. For example, the effects of the chestnut blight on the American chestnut and of Dutch elm disease on the American elm have been disastrous for these species though both continue to survive. It will be interesting to see whether they manage to adapt to these new selection pressures. Similarly, the introduction of such diseases as measles and smallpox, and more recently, AIDS, into previously unexposed human populations has had devastating effects and has put new selection pressures on these populations. These examples involving disease are striking and suggest that selection pressures are not static, but may be constantly shifting in subtle as well as not so subtle ways. Just as past bottlenecks in population size may be undetectable, the nature of past selection pressures may remain a matter of speculation. The cause of the extinction of the dinosaurs is a case in point, for theories abound, but the actual conditions that led to the downfall of the dinosaurs remain unknown. Whatever they were, however, it seems safe to state that the physical or biological environment changed in such a way that the dinosaurs were unable to adapt, which means that they were done in by a shift in selection pressure.

The crux of the matter is that the course of evolution may be influenced by chance events, that not all evolutionary changes are necessarily adaptive. The gene pool of a population may be buffeted by unpredictable mutations or recombinants or immigrants, or by random fluctuations in selection pressure or population size, all of which may contribute to the seeming stochastic nature of many evolutionary changes. The critics of the "adaptationist programme" accuse the adaptationists of ignoring the role played by these other factors or of merely paying lip service to them (Gould and Lewontin, 1979). However, the critics, in their attacks on the adaptationists, seem to err in the opposite direction in not recognizing that adaptation is an inescapable part of evolution. All organisms exist in a physical and biological environment, and thus are constantly subject to natural selection. For example, even when a population is small enough for genetic drift to occur, natural selection still continues to act in that population (Merrell, 1953a). The observed evolutionary patterns reflect the interplay between the adaptive and nonadaptive factors affecting the gene pool. Rather than evolution being either adaptive or nonadaptive, or directional or stochastic, it is both. The stochastic factors may come and go, however, but selection, in some form, is a constant presence.

In discussions of evolution, a stable environment is often postulated so that the selective forces can be assumed to be constant. Although this is a convenient simplifying assumption, it is unlikely to hold true for more than a very brief period. Even if a fixed, unchanging physical environment could be sustained, which is not the case even in the tropical forest or marine habitats usually regarded as highly stable, the species living in such environments will continue to evolve because the interactions among them will generate both intraspecific and interspecific selection pressures. Predation, parasitism, competition, and social interactions all continue to operate even under stable physical conditions. Thus Eldredge and Gould's idea (1972) that evolution is characterized by long periods of stasis interrupted by occasional bursts of rapid evolutionary change leading to speciation seems an unlikely scenario for the way evolution actually occurs. Even when stabilizing selection would seem to be the order of the day (for example, in the constancy of shell shape in the fossil record of certain molluscs), we know little about the physiological or biochemical changes that might have occurred within those shells during that time. The evolution of horses is well documented in the fossil record and is generally thought to have occurred in a stable grassland habitat, but nevertheless dramatic changes in size and morphology mark the evolution of the horses. It would take a bold person to state which of these changes was adaptive and which nonadaptive, but it does seem safe to say that there were some of both. In the final analysis, evolution is a composite of stochastic and directional changes, and superficial inspection and introspection are poor ways to distinguish one from the other.

Factors Enhancing Genetic Variability

Now it seems desirable to consider the shaping of the gene pool of a breeding population and the role genetic variability plays in evolution. We have already seen that natural selection is phenotypic, but that phenotypic selection will result in evolutionary change only if the phenotypic differences are to some extent hereditary. Moreover, the gene pool of a breeding population is the working material of evolution, and this pool must contain genetic variability, for no evolutionary change is possible in a completely homozygous population. If selection is to be effective in changing gene frequencies, alternative alleles must be available. Thus, genetic variability is the *sine qua non* of evolution.

First, let us consider the factors that maintain or increase the amount of genetic variability in a population. If mutation, selection, migration, and drift are absent, the existing genetic variability will be maintained in a stable Castle-Hardy-Weinberg equilibrium. Mutation tends to increase the amount of variability, as does an influx of genes from other populations.

Although natural selection is often thought to reduce genetic variability, under some circumstances selection will maintain it. Various types of balanced polymorphism are maintained by one or another form of balancing selection. For example, if a heterozygote is fitter than either of its corresponding homozygotes, selection will favor the fittest genotype, the heterozygote, and thus assure the continued presence of both alleles in the population. Similarly, frequency-dependent selection, where selective values shift as gene frequencies change, may also preserve variability. Opposing selection pressures in males and females as well as reversals in selection pressure during development will also permit the persistence of more than one allelic type. If selection is heterogeneous in time or in space (disruptive selection or the Ludwig effect), or if selection pressures are density related, these shifts in selection pressure may also preserve genetic variability. (For further discussion of balanced polymorphism, with examples, see Merrell, 1981, chapter 6.)

In addition to the types of balanced polymorphism cited earlier, selection may be balanced against other forces. For instance, if mutation pressure is opposed by selection pressure, with the mutation rate to deleterious mutants equal to their rate of elimination, a balance will be struck between mutation and selection, and of course, the same will hold for migration pressure if it is opposed by selection. In these cases, variability is preserved by the balance between opposing forces, with selection playing its more familiar role of eliminating the less favored types.

The mode of organization and transmission of the genetic material of a species is called its genetic system. The most familiar genetic system is that of a diploid species with separate sexes and homosequential chromosomes. The evolution of chromosomes and of mitosis and meiosis seems to have regularized the duplication, recombination, and distribution of the genetic material. In fact, the variety of genetic systems is most easily interpreted in terms of their effects on genetic recombination and the maintenance of genetic variability.

Sexual reproduction with the sexes separate ensures that outcrossing and genetic recombination will occur. A large number of relatively small chromosomes, high rates of crossing over, and an absence of chromosomal rearrangements all promote genetic recombination within a species. Other mechanisms that promote recombination or the maintenance of variability are negative assortative mating and various forms of enforced outcrossing such as self-sterility alleles or different maturation times for male and female gametes in hermaphrodites. Permanent hybridity may be ensured by systems of balanced lethals, or by inversion or translocation heterozygosity, or by allopolyploidy, each a device for maintaining variability and in many cases heterosis as well.

Factors Restricting Genetic Variability

Conversely, a number of mechanisms have been identified that lead to a decrease in recombination or variability. Asexual reproduction, which takes many forms, is the most effective of all, for it merely perpetuates existing genotypes. The only way evolution can occur in an asexually reproducing population is through the accumulation of sequential mutations. Inbreeding, and the ultimate in inbreeding, self-fertilization in hermaphrodites, will also reduce recombination and variability. Moreover, if the genetic material is organized into a small number of large chromosomes, and if the rate of chiasma formation and crossing over is low, genetic recombination will be minimized.

Other mechanisms reducing genetic recombination are the absence of crossing over in *Drosophila* males, which means that males can only transmit whole chromosomes, and heterozygosity for chromosomal rearrangements, which tends to lock up large intact blocks of genes within which no genetic recombination occurs in heterozygotes.

Among the other factors leading to a diminution in the amount of genetic variability, natural selection looms large. Already mentioned is its role in opposing the effects of mutation and migration by eliminating the less favorable phenotypes and their corresponding underlying genotypes, but directional selection and normalizing or stabilizing selection also tend to lead to a reduction in the amount of genetic variability. Directional selection is exemplified by transient polymorphism in which, if conditions change, a previously favored allele is superseded by a different allele more favorable under the new conditions. Stabilizing selection is also usually marked by a reduction in variability as the more extreme variants are eliminated along with the genes that make them poorly adapted to the existing conditions.

Random genetic drift in effectively small breeding populations also leads to the loss of alleles and a decrease in variability in natural populations. The subdivision of a species into a number of isolated breeding populations also causes a decrease in heterozygosity in the species (the Wahlund effect) compared with the amount available if all members of the species were interbreeding freely.

The picture that emerges is that of the gene pool of a breeding population in a state of tension between the forces that tend to increase the amount of variability in the population and those that tend to reduce it. If no variability were available, evolution would cease; if extremely high mutation, migration, and recombination rates prevailed, well-adapted individuals might become very rare. The genetic system itself is a product of evolution, and the wide variety of types of genetic systems suggests that there is no one genetic system that works best for all species. Rather it suggests that the genetic systems have evolved as mechanisms for regulating the amount of genetic variability and recombination in natural populations, and that the requirements differ among species.

The Evolution of Life Cycles and Genetic Systems

Let us consider briefly some of the possibilities. In some species, reproduction seems to be exclusively asexual whereas in others it is exclusively sexual, but in many species both asexual and sexual reproduction occur. In some of these cases, there is a regular alternation of sexual and asexual generations, but in others one or the other prevails most of the time until some special set of conditions triggers the other mode of reproduction. In aphids, for example, a series of asexual generations during the spring and summer months permits a rapid expansion of the population under favorable conditions. When conditions worsen in the fall, this triggers a sexual phase that generates a variety of new gene combinations just prior to the onset of winter. From this and other cases it has been argued that asexual reproduction is more efficient than sexual reproduction for the rapid increase of well-adapted genotypes in a stable environment. Asexual reproduction is also thought to be advantageous at low population densities when suitable mates may be rare or unavailable. Thus species with asexual reproduction are also well adapted to invade and colonize previously unoccupied but favorable habitats.

The essence of sexual reproduction is genetic recombination, and the continuous generation of new genotypes for testing against the existing conditions seems to be the main advantage of sexual reproduction. Especially among plants there are many species whose genetic systems include both sexual and asexual modes of reproduction so that they combine the best features of both modes of reproduction. Those species that can readily generate new genotypes through sexual reproduction and increase the frequency of the more favorable types rapidly through asexual means seem best able to weather the vicissitudes of life in a changing physical and biological environment.

The genetic system can be fine-tuned in a number of ways. (For more details see Merrell, 1981, pp. 402 ff.) Mutation rates are influenced by mutator genes, and selection for or against the presence of mutators in the population can thus regulate mutation rates. With sexual reproduction, inbreeding and selfing will reduce variability whereas outcrossing will help to maintain it. Thus, some species will be hermaphroditic, others will have separate sexes; some species will be self-sterile, others will be self-fertile (Grant, 1981). Furthermore, the amount of variability will be influenced by the size and number of the chromosomes, the frequency of chiasma formation and crossing over, and the presence or absence of chromosomal polymorphism. All of these phenomena are subject to selective pressure as well as to the effects of mutation, migration, and drift so that these characteristics, like other phenotypic traits considered earlier, are molded by evolutionary forces. In fact, the evolution of genetic systems represents a great challenge as well as a great opportunity for the student of evolution to gain greater insight into the inner workings of the mechanism of evolution.

In sexual reproduction, there are four fundamental processes involved: mitosis, meiosis, gametogenesis, and fertilization. There is also an alternation of haploid and diploid phases in the life history of the organisms. The sequence of the four fundamental processes determines the duration of the haploid and diploid phases. For instance, if meiosis immediately succeeds fertilization, the only diploid cell in the life cycle will be the zygote or fertilized egg, but if meiosis immediately precedes gametogenesis and fertilization, the gametes will be the only haploid cells in the life cycle.

Although a wide variety of life cycles has been observed in plants and animals, ranging across the spectrum between the two extremes, in general there has been an evolutionary trend toward a predominance of the diploid phase in the life cycles of both plants and animals. The usual explanation for this trend is that diploids can carry a reservoir of unexpressed variability in the heterozygous condition, which can be released through recombination each generation. In this way the population is able to respond to new selection pressures as they arise while remaining well adapted to the existing conditions. Moreover, diploidy permits interallelic, heterotic, and epistatic gene interactions that are not possible in haploids. Thus diploidy seems better able than haploidy to provide complex developmental controls as well as a means for the conservation and gradual release of genetic variability.

The DNA in the prokaryotes (viruses, bacteria, and blue-green algae) takes the form of a naked nucleic acid molecule whereas in the eukaryotes the DNA is associated with basic proteins to form more complex structures, the chromosomes, which are enclosed within a nucleus. Thus, the genetic system itself seems to have evolved from the relatively simple system seen in prokaryotes to the more complex system of the eukaryotes. The eukaryote genetic system makes possible a more orderly replication and distribution of the genetic material.

Although genetic recombination, and thus by definition sexual reproduction, occurs in the prokaryotes, their predominant mode of reproduction is asexual. The trend in the evolution of the higher or more complex life forms of plants and animals has been toward a predominance of sexual reproduction over asexual reproduction. The usual explanation for this situation is that more rapid evolution is possible with sexual than with asexual reproduction because sexual reproduction generates a greater variety of genotypes and phenotypes to be tested by natural selection. In quite a few higher forms, there has been a retreat from the cross-fertilizing, diploid sexual condition to some form of asexual reproduction, but even though such shifts may confer a short-term adaptive advantage, they seem to do so at the expense of long-range adaptability, for such groups appear to be evolutionary dead ends (Stebbins, 1950; Grant, 1981). Nevertheless, this retreat from sexuality suggests that under some circumstances asexual reproduction must have a selective advantage over sexual reproduction.

That the genetic system has adaptive significance is also indicated by the fact that genetic systems that restrict recombination occur in species in which immediate fitness and a high reproductive rate are at a premium. Genetic recombination can be limited in three ways: asexual reproduction, an increase in inbreeding, or limitations on the amount of crossing over. These mechanisms tend to be mutually exclusive; if inbreeding or selfing is the mechanism employed to reduce recombination, for example, asexual reproduction or limitations on crossing over will be rare in that group. This observation suggests again the opportunism of evolution and the combination of adaptive and chance nonadaptive features that mark evolutionary change, for where a reduction in genetic recombination appears to be favored by selection, the particular way this is achieved (asexuality, inbreeding, or crossover reduction) apparently depends on which mechanism selection is able to seize on at the time. This, too, of course, is a "Just So" story, but the genetic system is so crucial to the survival and reproduction of the members of a breeding population that it is difficult to escape the conclusion that it is subject to the same evolutionary factors that govern the evolution of the molecular, cellular, physiological, morphological, and other phenotypic characteristics of the individuals in the population. In truth, the principles that govern the evolution of genetic systems appear to be the same as those governing the evolution of DDT resistance in insects or any other type of evolutionary change. In other words, the trajectory of evolution results from the combined effects of both adaptive and stochastic events in a genetically variable breeding population.

Genetic Loads

If evolution is to occur, variability must be present, but the presence of variability appears to create a dilemma, for if the population consists of an array of genotypes and phenotypes, some will be better adapted than others, and the average fitness of the population will be lower than if it consisted only of individuals with the genotype best adapted to the existing conditions. This reduction in fitness owing to the presence of genetic variability is called the genetic load (Muller, 1950; Wallace, 1991). Various types of loads have been identified. (See Merrell, 1981, chapter 8 for more detail.) The input load refers to inferior alleles introduced into the gene pool by mutation or immigration. The balanced load is created by selection favoring allelic or genic combinations that by segregation and recombination form inferior genotypes every generation; the balanced load includes among others the segregational load, the recombinational load, the frequency-dependent load, the density-related load, and the loads related to heterogeneity of selection in time or space. The substitutional load, also called the cost of natural selection, is generated when selection favors the replacement of an existing allele by a new allele, which becomes the new "wild type" gene, most simply defined as the most frequent allele at a given locus.

Although the concept of genetic load has an appealing simplicity and is easily understood, it also leads into some logical tangles that limit its usefulness. For example, much of the discussion of genetic loads revolves around the question of whether the genetic load in natural populations is primarily mutational or segregational. However, it has been demonstrated (Merrell, 1981, pp. 190 f.) that in two populations so similar that it would be virtually impossible to measure any differences between them, the random mating load (L_r) would be 100 times greater if the load is assumed to be segregational than if it were mutational. Thus, in order to calculate the load properly, one requires prior knowledge of what type of load it is, which considerably limits the practical value of the concept. In these calculations, the gene and genotype frequencies are assumed to be in equilibrium, and the population size and the environment are assumed to be constant from generation to generation. All of these assumptions are very difficult to test and highly improbable.

The difficulties of the load concept are particularly obvious in relation to the substitutional load or the cost of evolution. When a population of insects is first exposed to DDT, large numbers of susceptible individuals die while a few resistant individuals survive. The cost is enormous, in terms of individual deaths, and the population size does not remain constant but declines precipitously. Nevertheless, the handful of survivors become the progenitors of the next generation, which will be more resistant to DDT than the previous generation. Thus the way evolution actually occurs is rather different from the way it is assumed to occur in estimates of the cost of evolution based on the genetic load concept, for Haldane (1957, 1960) and others have concluded that because of the inherent costs associated with allelic substitutions, evolutionary change can only proceed very slowly.

Another anomaly of the genetic load concept is that a homozygous population has no genetic load, but a single favorable mutation would immediately create a large genetic load, for all the other alleles would now be detrimental relative to the new mutant. The quickest way to eliminate this load would be to eliminate the favorable mutation, but that would not be the route to improved fitness. A genetically variable population will always have a genetic load, by definition, but it may well be thriving and better adapted and more adaptable than a homozygous population of the same species. The expression "genetic load" was coined by Muller (1950) to convey the idea that such a load is harmful to a population. To call the genetic variability in a population a genetic load is clearly misleading, however, for even though the variability may sometimes have harmful consequences, this so-called load is the genetic variability that forms the basis for the evolutionary potential of the population. Without genetic variability, evolution is impossible. Although the genetic load concept has been very useful in helping to clarify thought about the nature and effects of genetic polymorphism in natural populations, it appears that genetic load is a concept whose time has come, and gone.

Chapter 6
Gradual versus Saltational Evolution

The development of modern evolutionary theory has not been a steady advance from the natural selection of Darwin to the more complex modern synthesis of today, for progress has been accompanied by controversy at almost every step. Some of these controversies persist even today. One in particular is the argument over whether evolutionary change is gradual or saltational, continuous or discontinuous. This fundamental question has aroused strong opinions on both sides. However, when we examine the backgrounds of the partisans, we find a definite relationship between their position on the question and their previous training and experience.

Darwin, Huxley, and Galton

Charles Darwin, who first won wide acceptance for the concept of biological evolution and for natural selection as the mechanism of evolution, believed that evolution was a gradual process. He had the broadest interests and experience of any who have tackled this question and can best be described as a naturalist. His observations and collections during his five-year voyage on the *Beagle* brought him into contact with a wide diversity of living and fossil species, and his subsequent work on the biogeography, paleontology, morphology, embryology, and taxonomy of wild species as well as his studies of variation in domesticated species gave him a breadth of knowledge about species unequaled by that of his successors. Despite his achievements, it is worth remembering that Darwin was not infallible. He postulated sexual selection as an evolutionary factor separate from natural selection, although it is now clear that sexual selection is simply one of the components of reproductive fitness, the true measure of natural selection. In the successive editions of *On the Origin of Species* he became progressively more Lamarckian in his belief in the inheritance of acquired characteristics and even advanced a Lamarckian theory of heredity known as pangenesis. Both Lamarckism and pangenesis have long since been discredited.

Thomas Henry Huxley, known as Darwin's bulldog for his vigorous support of Darwin's ideas, did not follow Darwin blindly, for he believed in discontinuous evolution by means of "sports" or saltations as did Darwin's cousin, Sir Francis Galton. Huxley, the son of a poor schoolmaster, passed his medical exams and became a surgeon in the English fleet exploring the northern coast of Australia. He became fascinated by and began to study the many exotic species in those tropical waters. He eventually published several biological works that brought him recognition and a professorship in England, where he continued his research and teaching in physiology, comparative anatomy, and paleontology. Huxley's support of discontinuous evolution was based in part on his familiarity with the short-legged Ancon breed of sheep and with polydactyly in humans. Both are simply inherited, stable traits without intermediates and seemed to provide a mechanism for evolution by saltation.

In his youth Galton began to study medicine, but when he inherited his father's fortune, he left school and set off on journeys to Egypt and southwest Africa, where he made notable geographical discoveries. On his return to England he worked on meteorology for a few years but then turned to the subject that became his lifelong vocation, the study of heredity. His first efforts were designed to test Darwin's theory of pangenesis experimentally. His failure to confirm the existence of pangenes led him to oppose the inheritance of acquired characteristics and to develop his own theory of heredity, which involved "stirps" as the hereditary material. This theory also turned out to be invalid, and Galton went on to develop the statistical approach to heredity for which he is best remembered today. However, Galton, like Huxley, believed in the stability of sports and for many years stressed the possibility of discontinuous evolution based on discontinuous variation. His failure to confirm pangenesis may have made him equally skeptical of Darwin's concept of gradual evolution or else his work with human heredity may have drawn his attention to discontinuous variation as the basis for evolution.

Biometricians versus Mendelians

Oddly enough, Galton's closest associates and successors in the development of biometry, W. F. R. Weldon and Karl Pearson, both supported the concept of continuous evolution. Weldon was trained as a morphologist and embryologist, but later became interested in a statistical approach to variation and evolution although he never became an outstanding statistician. Pearson, on the other hand, had a remarkable background. He graduated from Cambridge with honors in mathematics, but there and subsequently in Germany studied philosophy, physics, engineering, law, medieval languages, metaphysics, and even Darwinism. He was admitted to the bar but wound up at age twenty-seven in the chair of applied mathematics and mechanics at University College, London. Pearson was influenced by Galton's book *Natural Inheritance,* but his interest in evolution, heredity, and statistics was greatly stimulated by Weldon, who felt the need for assistance from someone with Pearson's mathematical skills. Their statistical methods were particularly well suited for the study of continuous variation, so that it is not surprising that they not only studied continuous variation but believed that it was the basis for gradual, continuous evolutionary change.

The hostility between the biometricians, Weldon and Pearson, and the Mendelians, particularly William Bateson, arose not only from their differences over the mechanism of heredity, but from their disagreement over whether evolution was continuous or discontinuous. In 1894 Bateson had published a major work called *Materials for the Study of Variation* in which he marshaled evidence in support of the idea that discontinuous variation was the basis for discontinuity in the origin of species. The rediscovery of Mendelism in 1900 merely strengthened this belief, in which he was soon joined by two eminent geneticists, Hugo De Vries and Thomas Hunt Morgan.

Although Bateson was originally trained in morphology and embryology and was never very competent in mathematics, a handicap in his disputes with the biometricians, he soon turned his attention to questions of variation and evolution. With the rediscovery of Mendelism in 1900, he became one of its most ardent advocates, and felt that it vindicated his belief in discontinuous variation as the basis for evolutionary change. De Vries, one of the codiscoverers of Mendelism, joined him in this belief with his work *The Mutation Theory* (1901–3) for he had discovered what appeared to be new species in the evening primrose, *Oenothera,* that arose at a single step rather than gradually. Morgan's original interests were in embryology, but he turned to the genetics of *Drosophila* in the early days of this century, laying the foundations for the discovery of linkage and crossing over, work for which he received the first Nobel Prize awarded to a geneticist. Bateson, De Vries, and Morgan shared not only the fact that they were Mendelians but also that they were experimentalists; all were involved in breeding experiments with plants or animals or both. In their experiments geneticists usually seek out clear-cut, well-defined variants because the crosses are then much easier to monitor and to score. Perhaps it should not be surprising that those who worked daily with discontinuous variation came to regard it as the basis for evolution.

However, another group of experimental geneticists supported gradual rather than saltational evolutionary change. William E. Castle, H. Nilsson-Ehle, and E. M. East were all Mendelians, and all eventually wound up supporting the concept of gradual evolutionary change. Castle, trained before 1900 as an embryologist, became one of the leading geneticists in the United States in the first decade of this century. Like Bateson he was not well trained in mathematics, and thus was poorly equipped to tangle with the biometricians. Although Castle at first accepted discontinuous evolution based on De Vries's mutation theory, the results of his selection experiments on quantitative traits soon caused him to change his mind. East, Castle's colleague at Harvard, was originally trained as a chemist but became involved in plant breeding experiments on the oil and protein content of corn; he went on to a career as a plant breeder, which was also the background of Nilsson-Ehle, a cereal breeder in Sweden. Castle, East, and Nilsson-Ehle all demonstrated that selection could be effective on quantitative traits, and from their work came the Mendelian multiple factor hypothesis for the inheritance of quantitative characters and its corollary, the concept of modifying factors. Again the nature of their research seems to have played a crucial role in the position they took with respect to continuous versus discontinuous evolution.

The Population Geneticists

R. A. Fisher, J. B. S. Haldane, and Sewall Wright, the three people primarily responsible for the development of population genetics during the 1920s and early 1930s, were quite diverse in training and background. Fisher, like Pearson, was trained primarily as a mathematician, but unlike Pearson, when Fisher became interested in heredity, he was soon convinced of the validity of Mendelian genetics. He suffered from severe myopia, a considerable handicap, it would seem, for one who was also a student of astronomy. Like Pearson he was a brilliant student, spending an extra year at Cambridge after graduation studying physics, but subsequently his interests were focused on mathematical statistics and evolutionary theory. His greatest contribution to evolutionary theory was to synthesize the concepts of Darwinism, Mendelism, and biometry into one consistent and coherent pattern. Because his vision was so poor, he did little experimental work, but instead analyzed the data of others from a variety of animal and plant species.

Haldane, like Fisher and Pearson, was a brilliant student. Son of the well-known physiologist J. S. Haldane, J. B. S. was exposed to scientific research at an early age. He even started doing genetic research on linkage in small mammals before going to college. He entered Oxford on a mathematics scholarship, winning first-class honors within a year, but then switched to philosophy and classics. Although he did study biochemistry later with Gowland Hopkins at Cambridge, his formal training in biology was limited. His interest in population genetics was sparked by the mathematical problem of quantifying the effects of natural selection in populations. Despite his early exposure to experimental research, Haldane, like Fisher, spent most of his career analyzing the data of others. One story, possibly apocryphal, is that he tried to learn to work with *Drosophila*, but he tended to snort at inopportune times and blow away the flies from under the microscope, so he gave it up as a bad job.

Both Fisher and Haldane were strongly selectionist in their outlook. Fisher came closer to the Darwinian position because he believed that evolutionary change resulted from the effects of mass selection on numerous small genetic differences in large populations. Haldane also believed in the overriding importance of natural selection but thought that the selection of genes of major effect could also be of considerable importance in producing rapid evolutionary change.

Unlike Fisher and Haldane, Wright had relatively little formal training in mathematics although he obviously had a knack for it and taught himself a great deal of math as the need arose in his work. He started studying biology at a small college in Illinois, and eventually wound up at Harvard studying genetics with Castle. His initial research was in physiological genetics, a study of the inheritance of coat color and other traits in guinea pigs. Wright also helped Castle with the experiments on hooded rats that first demonstrated the existence of modifying factors. For the rest of his career, Wright was involved in breeding experiments, especially with a long-term study on the effects of inbreeding in guinea pigs. His shifting balance theory of evolution, based on natural selection acting on interacting systems of genes generated by mutation, migration, recombination, and random drift in relatively small populations, undoubtedly stemmed from his observations and experiments on his inbred populations of guinea pigs. Although more complex than Fisher's or Haldane's ideas about how evolution occurred, Wright's theory also falls into the category of a mechanism for gradual evolution.

The Modern Synthesists

The three best-known advocates of the modern synthesis, Theodosius Dobzhansky, Ernst Mayr, and G. G. Simpson, had backgrounds rather different from those of Fisher, Haldane, and Wright. Dobzhansky's early work as a field naturalist was focused on the distribution and systematics of ladybird beetles of the family Coccinellidae. Although he later became expert on the genetics and cytogenetics of *Drosophila* under the tutelage of A. H. Sturtevant, he continued to rely on Wright and others to help him with the mathematical aspects of his research in population genetics. Mayr was trained as an ornithologist and spent the early years of his career working on the systematics of the birds of the South Pacific. Interestingly enough, he was a neo-Lamarckian in the 1920s but had become a neo-Darwinian by the mid-1930s. Although he did some experimental work on behavior in *Drosophila* in collaboration with Dobzhansky, he hardly qualified as an experimentalist nor did he have much training or experience in genetics or mathematics. Both Mayr and Dobzhansky believed that the gradual accumulation of genetic differences in different populations of the same species eventually led to the origin of species. The genetic divergence is primarily the result of natural selection, which produces populations well adapted to their respective environments. Mayr was particularly vehement in his support of allopatric or geographical speciation.

Simpson was also, in a sense, trained as a systematist, but his research dealt with vertebrate fossils rather than with living species. Nevertheless, he seemed more comfortable with the theoretical genetic concepts advanced by Fisher, Haldane, and Wright than did Dobzhansky or Mayr. He also seemed more comfortable with math, for in collaboration with his wife, Anne Roe, and later Richard Lewontin, he wrote a very useful statistics book entitled *Quantitative Zoology* (1960). Although he identified three modes of evolution in his 1944 classic *Tempo and Mode in Evolution* (phyletic evolution, speciation, and quantum evolution), nonetheless all three modes meet the criteria for gradual evolutionary change though some modes are more rapid than others. Simpson had to confront the problem of gaps in the fossil record, but he never resorted to saltational evolution as an explanation.

Thus, during the 1930s and 1940s a consensus developed among the most influential biologists of the day that evolution resulted from the gradual accumulation of numerous genetic differences among different breeding populations of the same species. Darwinian selection was reconciled with Mendelian genetics with a few added frills such as mutation, migration, and drift.

Goldschmidt and Gould

About the only apostate was Richard B. Goldschmidt. Goldschmidt's primary training was in cytology, but he had a broad classical training in zoology in such disciplines as morphology and embryology before he gravitated toward genetics. His original interests lay in physiological and developmental genetics, areas in which he did extensive research on the phenogenetics and sex determination of *Drosophila* and the gypsy moth. In this work he also studied the numerous geographical races of gypsy moths. Rather than accept the current evolutionary dogma, which soon came to be called the modern synthesis, in 1940 Goldschmidt proposed, in *The Material Basis of Evolution*, that the ideas embodied in the modern synthesis were adequate to explain microevolution, the origin of ecological races and geographical subspecies, but that for the origin of species, or macroevolution, other processes were required. His research on developmental genetics involved working with homeotic mutants and other genes of major effect, so perhaps it is not surprising that he postulated that macroevolution resulted from macromutations giving rise to hopeful monsters, a new version of discontinuous or saltational evolution. Goldschmidt was not mathematically adept, however, and did not really address the question of how these macromutations could become established in the gene pool of a population. For this reason, his ideas did not win wide acceptance.

More recently, Niles Eldredge and Stephen J. Gould (1972) have postulated what they consider to be a new version of discontinuous evolution, which they call "punctuated equilibrium." Gould has also attempted to resurrect Goldschmidt's ideas by reissuing his *Material Basis of Evolution* in 1982. Eldredge and Gould are primarily invertebrate paleontologists and systematists rather than experimentalists, and their concept of discontinuous or saltational evolution is fitted to a rather different time scale from that used by a neontologist. Once this is realized, it is clear that Eldredge and Gould's "punctuated equilibrium" is more or less equivalent to Simpson's "quantum evolution," and that they conceive of punctuated equilibrium as occurring in tens or hundreds of thousands rather than millions of years. Evolution over that span of time, however, fits a gradual pattern rather than an abrupt saltational mode of evolution as it is usually envisioned. For example, artificial selection in horses over the past 10,000 years has produced animals as diverse as Percherons and Shetland ponies. These changes are usually regarded as gradual, but if found in the fossil record over such a short period of time, they would undoubtedly be called saltational.

Present Views

If one reads such current mathematical theorists as J. F. Crow, M. Kimura, D. Hartl, R. Lande, and R. C. Lewontin, it is clear that although they may differ as to details and the relative emphasis they place on different factors, they are generally agreed that evolutionary change results from the gradual accumulation of numerous genetic differences among populations, primarily the effect of natural selection acting on quantitative traits, though Kimura, more than the others, has stressed the importance of neutral alleles in evolution. Crow, Lewontin, and Hartl are experimentalists as well as theorists, working primarily with *Drosophila*, whereas Kimura and Lande, like Fisher and Haldane, have worked mostly with the data of others.

This brief recital of the controversy over continuous versus discontinuous evolutionary change should suffice to provide some feeling for the history of the controversy. Over the years there have been some highly respected and knowledgeable biologists on both sides. The coming of Mendelism at first seemed to lead to greater strength for the saltationists, but more recently, gradualism seems to have won the day. The position taken often seemed to be related, not to whether the biologists were naturalists or experimentalists or theoreticians, or geneticists, ecologists, systematists, paleontologists, or mathematicians, but rather to the nature of the traits and organisms most familiar to them. Even more interesting is that the dispute always seemed to be framed in terms of alternatives: evolution was either continuous or discontinuous, gradual or saltational. If evolution could be shown to be gradual, then saltational evolution must be in error. There seems to have been a search for a universal mode of evolution, and that mode, according to the modern synthesis, is the allopatric gradual origin of species.

That people should have strained so hard to come up with a single mode for the origin of species is rather puzzling in view of the unrecorded millions of species of plants, animals, and micro-organisms that have existed. Given the variety of genetic systems, life histories, and environments of these species, it seems highly improbable that all would have evolved in a similar way. Perhaps a simple example will suffice to demonstrate that in fact they have not. Although exact estimates are difficult to obtain, it is widely agreed that among the higher plants from one-third to one-half or more of all the species are allopolyploids (Grant, 1981). Allopolyploidy is the result of hybridization between two different species followed by a doubling of the chromosome number. The product is a fertile new species, reproductively isolated by sterility barriers from its two parental species. Moreover, such species arise by sympatric rather than allopatric speciation and originate at a single step rather than gradually. Since at least a third of all higher plant species originate sympatrically and saltationally, this can hardly be regarded as some rare aberrant mode of speciation, and this example in itself should suffice to lay to rest the idea that the origin of species requires the gradual accumulation of numerous small genetic differences in allopatric populations. Once we are rid of this shibboleth, perhaps it will be possible to take a new look at the ways that evolution can occur and new species originate.

Chapter 7
Evolutionary Statics

Evolution is a historical process and is the composite result of adaptive and nonadaptive changes in a breeding population, the basic evolving unit. The mechanism that drives evolution consists of four independent factors: mutation, natural selection, random genetic drift, and gene flow. The relative importance of these four factors may vary in different populations and even in the same population as conditions change. The actual course of evolution may reflect the effects of both directional and random events. The species, which is the largest and most inclusive breeding population, consists of a number of demes tied together by migration and gene flow into a Mendelian population sharing bonds of parentage and mating.

Two fundamental problems of evolutionary biology are, first, to explain phyletic evolution (anagenesis), the evolutionary changes occurring within a single line of descent, and second, to explain speciation (cladogenesis), the splitting of a single species into two or more separate, reproductively isolated, contemporary species. Speciation is a more difficult problem than phyletic evolution, but both types of evolution can be satisfactorily explained within the framework of the modern synthesis as outlined earlier. A different explanation has sometimes been sought for the origin of families, orders, classes, and so on, but the unit that evolves is always a breeding population, and the evolutionary changes giving rise to these higher taxonomic groups, no matter how bizarre the final outcome compared to the ancestral population, have taken place in a succession of breeding populations, and are still the product of the combined effects of mutation, selection, migration, and drift.

Measuring Variation

We need to scrutinize this evolutionary mechanism more closely in order to understand how it actually works. Phyletic evolution or anagenesis occurs when a population with one set of characteristics gives rise to a population directly descended from it with a somewhat different set of characteristics; that is, it involves descent with modification. But what is it that changes, what characteristics should we examine? The answer is that many things may change. Phenotypes may change, but they are notoriously easily influenced by the environment. Genotypes may change, but we have already pointed out the difficulties of determining the complete genotypes of organisms. Allelic frequencies may change or allelic substitutions may occur, but in order to detect such changes, we must first know what alleles are present and what their frequencies are. Thus the first problem is one of evolutionary statics. Before it is possible to measure change, we first need to know something about the characteristics of existing populations. This has led to what is sometimes referred to as the "struggle to measure variation" (Lewontin, 1974).

In their efforts to study evolution, biologists originally compared the morphological characteristics of different groups. The variations observed became the basis for distinguishing among different subspecies, species, and other taxonomic groups, and much taxonomic work is still based primarily on morphological traits.

When geneticists started to assess genetic variability in natural populations, their first efforts were based on studies of visible mutants, those with an easily discernible effect on the phenotype, but these were relatively rare in wild populations. The next studies, mostly with *Drosophila*, determined the frequencies of lethal and semilethal genes in natural populations. The results showed that from one-quarter to one-half of the major chromosomes in most species carried at least one lethal or semilethal mutation. This finding came as quite a surprise and has some rather significant implications, but it was difficult to believe that lethals and semilethals could play a significant role in evolution. Adaptive evolutionary changes could hardly be based on a progression or succession of harmful mutations, most of them detrimental under any environmental conditions. Thus even though great efforts were devoted to the determination of lethal frequencies in natural populations, the data provided little insight into the genetic basis for evolutionary change.

Similarly, considerable work was done, especially by Dobzhansky (1970), on the role of inversion polymorphism in wild *Drosophila* populations, but apart from the finding that the inversions usually ensured permanent heterosis, no consistent role for all this chromosomal variation was discovered (Merrell, 1981, chapter 9). However, the discovery of so much chromosomal polymorphism did undermine the prevailing conventional wisdom that the gene pool of a species is bound up in groups of homosequential chromosomes. Subsequently, the finding of chromosomal polymorphism in many other groups came as less of a surprise, but its origin and role in natural populations continue to be problematic.

The use of gel electrophoresis for the study of protein variation marked the advent of a new era in the struggle to measure variation (Lewontin, 1974). Rather than looking at morphological variation or searching for visible or lethal mutations, biologists studied the migration of proteins in an electric field. Because each protein molecule has a characteristic charge and configuration, it migrates at a characteristic rate in the field, and after suitable staining, an array of proteins is revealed as a series of bands in the gel. This sensitive technique permits comparisons of the proteins in different organisms, although not all variants are detected because not all base substitutions cause a change in migration rate. Nonetheless, gel electrophoresis marked a considerable advance over previous methods of studying variation because it was possible to study genetic variation in natural populations at the level of primary gene action without significant environmental distortion. Furthermore, the proteins could be detected whether segregating alleles were present or not, so that it was possible to estimate the numbers of invariant as well as variable genetic loci.

The biggest drawback to the new approach was that in many cases the function of the proteins in the organism was understood very poorly or not at all. Thus, it was difficult even to speculate about, let alone test, their possible adaptive significance. Nevertheless, the new approach led to a burst of research activity that started over two decades ago and has not yet abated. The dearth of knowledge about the functional or adaptive significance of the proteins undoubtedly helped to foster the notion that these protein polymorphisms were adaptively neutral, and the controversy over neutralism versus selectionism reached its greatest intensity following the discovery of high levels of electromorphic variation in natural populations (Merrell, 1981, chapter 13). Another factor favoring the adoption of neutralism was that so much protein polymorphism was revealed that the then popular theories for explaining genetic polymorphism as the result of selective forces seemed inadequate to account for the tremendous amounts of protein polymorphism being revealed. Neutralism provided a very easy way out of the apparent dilemma: simply declare the polymorphisms adaptively neutral and the dilemma vanished. Since, by the very nature of scientific proof, it is impossible to prove adaptive neutrality, the problem was neatly laid to rest. Because the selectionists, often lacking clues as to the functional role of the alternative protein types, had difficulty testing the adaptive significance of the different electromorphs, they too lacked evidence to support their position. The dispute thus degenerated into a conflict between two different articles of faith, the neutralists firmly believing that protein polymorphisms are selectively neutral, and the selectionists that they are maintained by selective forces. The debate finally seemed to reach a stalemate and most of the participants declared a truce though some were still willing to declare victory by fiat. However, studies to determine the adaptive significance of protein variations have begun to reveal significant selective effects, and the picture is gradually changing (Nevo, Beiles, and Ben Shlomo, 1984; Nevo, 1988; Gillespie, 1991).

Gel electrophoresis is still widely used to study genetic variation, but it suffers from the drawback that it cannot detect all the variants. Estimates are that as few as a third of the variants are detected by the usual electrophoretic techniques, so that the amount of variation present will be underestimated unless special precautions are taken. Fortunately, other approaches are available to study molecular variation though they are often more time-consuming and expensive than gel electrophoresis. Among them are the amino acid sequencing of proteins, base sequencing of DNA and RNA, DNA and RNA hybridization, and more recently, the study of mitochondrial and plastid DNA.

The study of genetic variation in natural populations had one consistent result: the amount of variation revealed always seemed to be far greater than most workers expected. The preconception of a species composed of a number of phenotypically and genetically similar individuals was gradually eroded to the point where a new paradigm had to be sought. Not only genic but chromosomal polymorphism was rampant. The old idea that all the members of a species were homozygous for the "wild type" allele at nearly every locus and that these loci were similarly distributed among a group of homosequential chromosomes was clearly false. In a natural population the proportion of polymorphic loci may range from about 15 percent to more than 50 percent and the proportion of heterozygous loci per diploid individual from about 4 percent to more than 15 percent. Although

comparable reliable data are not available for the amounts of chromosomal polymorphism in wild populations, the information on inversion polymorphism in *Drosophila* and on chromosomal polymorphisms in small mammals suggests that chromosomal polymorphism
in natural populations, like genic polymorphism, is the rule rather
than the exception. Furthermore, White (1978) estimates that only
about 1 percent of closely related species share homosequential karyotypes, with 10 percent the absolute maximum, so that chromosomal
evolution as well as genic evolution seems to be the normal accompa-
niment to the origin of species rather than a rare aberration.

Genetic Polymorphism

One of the major revelations from the study of genetic variability in
wild populations was the high level of genic polymorphism. The
problem then became to explain why the level was so high and how it
was maintained at this level. The measurements were usually made at
a single point in time, and one of the questions that had to be addressed was whether the variability was transient or in equilibrium.
Repeated long-term sampling is necessary to establish trends in gene
frequency change and thus to demonstrate the existence of transient
polymorphism. Therefore, it has almost always been assumed that
the observed high levels of genic polymorphism represent equilibrium values, and the question then became what types of balanced
equilibria can sustain these high values.

The following are some of the possibilities. Because the vast majority of spontaneous mutations are deleterious (Simmons and Crow,
1977) and because natural selection, given the opportunity, will tend
to eliminate deleterious mutations from the population whenever
they are expressed in the phenotype, mutation pressure and selection
pressure always tend to be opposed to one another. If all deleterious
mutations were expressed as soon as they occurred, their elimination
would be very rapid, and the level of genic polymorphism from this
source would be low. Many deleterious mutations are recessive or
nearly so (Mukai et al., 1972), however, and their rate of elimination is
much slower than if they were dominant. Consequently, the balance
between mutation and selection can be a major contributor to the observed genic polymorphism.

Another possibility is a balance between migration pressure and
selection pressure. If immigrants introduce less-favorable alleles into
a well-adapted population, natural selection will tend to eliminate
these deleterious alleles just as it does those introduced by mutation.

Under most circumstances, it is extremely difficult to determine migration rates and the amounts of gene flow actually taking place unless the movement of individuals can be followed and their genetic contributions to the new population monitored. Because of these difficulties, the potential contribution of a migration-selection balance to any observed genic polymorphism tends to be ignored or dismissed, but its possible significance should not be overlooked.

Ford (1940) originally wrote, "Genetic polymorphism is the occurrence together in the same locality of two or more discontinuous forms of a species in such proportions that the rarest of them cannot be maintained merely by recurrent mutation." It is clear that his intent was to restrict the definition of genetic polymorphism to situations resulting from a balance between opposing selective forces within a single breeding population and to exclude from his definition transient polymorphism as well as the balance between mutation and selection or between migration and selection. However, because mutation rates and selection and migration coefficients are so rarely known, this distinction is usually impossible to make, and a polymorphic locus is best defined simply as one at which the frequency of the most common allele does not exceed 99 percent (Merrell, 1981, p. 48). Ford's effort to restrict the definition of balanced polymorphism to the genetic polymorphisms based on opposing selective forces, and to exclude transient polymorphism and the mutation-selection and migration-selection equilibria, has remained conceptually useful.

Balanced Polymorphism

The most familiar explanation for balanced polymorphism is based on heterozygous advantage where the heterozygote for an allelic or chromosomal difference is superior in fitness to both of the corresponding homozygotes. At one time heterozygous advantage was accepted as the explanation for genic polymorphism almost to the exclusion of all other possibilities. For instance, a dominant gene known as Burnsi is found in up to 10 percent of the frogs in wild populations of leopard frogs over thousands of square miles in Minnesota and adjoining states. A well-known ecological geneticist, shown some of the unspotted Burnsi frogs in a holding tank and given a brief explanation of their distribution and frequency in natural populations, instantly declared that heterozygous advantage must be maintaining the gene in the wild populations. Unfortunately, no evidence for heterozygous advantage ever turned up (Merrell, 1972), although seasonal selection was demonstrated (Merrell and Rodell, 1968).

Heterozygous advantage has always been a popular explanation for balanced polymorphisms, in large measure, perhaps, because of its appealing simplicity. The usual procedure is to compare the relative values of one or another of the components of fitness such as viability or fecundity for the three genotypes, and if the heterozygote is superior to the homozygotes in this test, it is often assumed to be superior to the homozygotes under all conditions. Thus, the need to assess the three genotypes relative to differing environmental conditions is eliminated, and the experimental procedure is greatly simplified.

The discovery of ever-increasing amounts of genetic variability in natural populations finally led to a decline in the popularity of heterozygous advantage as an explanation for genic polymorphism. The reason was that with heterozygous advantage, two less fit homozygous types will always be formed along with the favored heterozygotes. A population with relatively few polymorphic loci may be able to tolerate the presence of such homozygotes, sometimes referred to as the segregational load, but as the number of polymorphic loci increases, practically every individual in the population will be homozygous at one or more of these loci, and the segregational load would seem to become intolerable. In an effort to surmount this problem, King (1967), Milkman (1967), and Sved, Reed, and Bodmer (1967) independently pointed out that some form of threshold model with truncation selection could help to decrease the segregational load (Wills, 1981).

It is worth noting that it is extremely difficult experimentally to perform an adequate test of single-locus heterosis, that is, to demonstrate that an observed heterozygous advantage is owing to the interaction between alleles at a single locus. Usually the observed heterosis is due to the interactions between the genes on two small chromosome segments; that is, it is probably due to dominance and epistasis rather than overdominance as originally defined. Furthermore, the most rigorous tests for interallelic overdominance (Simmons, 1976; Falconer, 1981) have rarely demonstrated its presence. Therefore, it seems likely that heterozygous advantage based on single-locus heterosis is far less common than was formerly believed and can account for at best only a very small fraction of the genic polymorphism found in wild populations. The lack of convincing evidence for single-locus heterosis and the large segregational load that would be generated if it were true both suggest that interallelic overdominance is a poor candidate to explain the high levels of genic polymorphism.

When heterozygous advantage began to lose its appeal, other explanations were sought for balanced genic polymorphism. One of the more popular was frequency-dependent selection, which can take several forms. In essence, this theory postulates that the selective values of the genes are not constant but change as the gene frequencies change. If selection against the rarer type becomes stronger the rarer it gets, its rate of elimination will be hastened. However, if the rarer allele or genotype is always favored by natural selection, a stable equilibrium will be reached at the point on the frequency curve where the various types are equal in fitness. If this equilibrium is perturbed, the rarer type will always be favored by selection, and the population will tend to return to the equilibrium point. This theory has considerable appeal as an explanation of balanced polymorphism because, at equilibrium, all the genotypes have equal fitness and there is no genetic load comparable to the one generated by the less fit homozygotes under the theory of heterozygote advantage.

Despite its theoretical appeal, the experimental evidence for frequency-dependent selection is somewhat less than satisfactory (Kimura, 1983). Studies of enzyme polymorphism, for example, have been reported to demonstrate frequency-dependent selection with a rare type advantage in some cases but not in others, but statistical and methodological pitfalls abound that have not always been avoided. Similarly, a great amount of work has been done on rare male mating advantage in *Drosophila* where the experimental and statistical and interpretive pitfalls are even greater. My own experience with mating experiments with the white-eye mutant and its wild-type allele in *Drosophila melanogaster,* in which the wild-type males showed a rare male mating advantage but the white-eyed males showed a rare male mating disadvantage, has made me wary of simple interpretations of mating experiments (Peterson and Merrell, 1983). This situation, rather than generating a balanced polymorphism, leads to a more rapid elimination of the deleterious white gene, and is in accord with expectations based on a knowledge of the biological effects of the alleles.

One of the difficulties with postulating a rare type advantage is that, apart from its usefulness to the theory, little thought is usually given as to why the rarer type should always be advantageous. There are, however, some situations in which the rationale for rare type advantage is obvious. For example, in many plant species and some animals self-incompatibility systems of alleles have evolved that prevent self-fertilization and close inbreeding. A pollen grain carrying a particular allele will not grow on the stigma of any plant carrying that allele. Thus, the rarer an allele is, the more potential mates it has in the population, but as it becomes more common, this selective advantage is lost. Therefore, self-sterility alleles provide a clear-cut case of balanced genic polymorphism based on frequency-dependent selection with the rarer types of alleles always favored.

Another situation in which the rationale for frequency-dependent selection seems equally obvious involves predator-prey relations. In hunting, many predators form search images of their prey to enable them to find prey more efficiently. In polymorphic populations, the search image will most often center on the most common type of individual, and these individuals will suffer disproportionately. As their numbers dwindle, the predators will focus their attention on some previously rare but now more common type, and a balanced polymorphism will result. Although this case is logically satisfying, well-documented examples are apparently rare (see, e.g., Endler, 1986).

Other theoretical possibilities for the maintenance of balanced genic polymorphism can be suggested. Negative assortative mating, for example, would ensure balanced polymorphism, but despite the adage that opposites attract, there is little evidence in support of this type of mating behavior, and negative assortative mating seems unlikely as a source of balanced polymorphism. Quite a few instances of "one-sided mating preferences" have been reported where one male type, for example, is more successful in mating than the other, and of course, sexual isolation results in positive assortative mating, but neither of these will generate balanced polymorphism.

Other types of opposing selection pressures that could maintain balanced genic polymorphisms may be cited. Selection might, for instance, favor one allele in males and another in females, or one allele at one stage of development and another allele at another. Both of these are probably overly simplistic, and not much supporting evidence is available. In cases of meiotic drive, in which a heterozygote produces an excess of one of the two possible types of gametes, the driven allele seems to be invariably associated with a deleterious effect at some other stage of the life cycle, so that the net effect is a balanced polymorphism.

Heterogeneous Selection Pressures

Another possible source of polymorphism is the Ludwig effect, which is similar in some ways to the balance between migration and selection discussed previously. This effect arises if the environment is heterogeneous in space (Levene, 1953; Hedrick, Ginevan, and Ewing, 1976). In such a situation, one allele or genotype may be favored in one subniche or microhabitat and others in other subniches or microhabitats. If mating is more or less at random in the area or if significant gene flow occurs among the various subniches, then a balanced polymorphism could exist. However, the difficulty in determining mating patterns in the wild and in measuring the amount of gene flow in natural populations has inhibited efforts to demonstrate this sort of polymorphism.

If selection pressures are heterogeneous through time, so that there are temporal or seasonal or cyclical changes in selection pressure with one allele or genotype being favored at one time and others at other times, these opposing pressures could support balanced polymorphism. Similarly, if fitnesses change with shifting population densities, density-related balanced polymorphisms could be generated. Temporal heterogeneity and density-related shifts in selection pressure have often been dismissed as sources of balanced polymorphism because theoretical considerations led to the conclusion that they would produce unstable equilibria and eventual fixation rather than balanced polymorphism (Kimura, 1955).

This conclusion was usually based, however, on the assumption of random changes in density or random temporal shifts in selection pressure. If the changes in density were cyclical or the changes in selection pressure were seasonal, for example, the polymorphisms generated in this way, though still inherently unstable, would be much more likely to persist and, from a practical standpoint, might be impossible to distinguish from a stable balanced polymorphism (Merrell, 1981, pp. 170–76; Ginsburg, 1983). Therefore, these possibilities cannot very well be ignored and should be added to the arsenal of possible explanations for genic polymorphisms in natural populations.

The discovery of high levels of genetic polymorphism and of heterozygosity in natural populations has had certain consequences. First, it was realized that heterozygous advantage, formerly by far the most popular explanation for balanced polymorphism, could account for only a small fraction of the variability at best, and other explanations were needed. Frequency-dependent selection, which got rid of the segregational-load problem inherent in heterozygous advantage, next became popular, but the evidence in its favor was often weak. Moreover, other than to meet the demands of the theory, a biologically convincing rationale for the rare type always to have a selective advantage was generally lacking. The usual inability to distinguish between transient and balanced polymorphisms led to the assumption that the observed polymorphisms were balanced, but without better knowledge of just how stable the physical and biological environment is, this remains merely an assumption. There is little evidence to show what proportion of genic polymorphism is balanced and what proportion is transient.

In addition to heterozygous advantage, frequency-dependent selection, and transient polymorphism, other mechanisms can also be responsible for the observed genic polymorphism. Among them are density-related selection, selection heterogeneous in time or space, and the balance between mutation and selection or between migration and selection. The problem is not that we lack explanations, but rather that there are too many, with no easy way to distinguish among them. In order to solve the puzzles of polymorphism, we need detailed knowledge of the species and its biological and physical environment; in other words, we need to study the ecological genetics of the species.

Chapter 8
Evolutionary Dynamics

In the preceding chapter we dealt with the struggle to measure variation. In this chapter we shall deal with the struggle to study evolution. After centuries of belief in the fixity of species, scientists began to realize that species are not fixed but ephemeral and constantly subject to change. Darwin, more than anyone else, summarized the evidence and convinced the scientific world that evolution had occurred. He drew evidence from many sources: domesticated species, the fossil record, biogeography, systematics, morphology, and embryology.

Although Darwin won acceptance for evolution, his efforts to establish the validity of natural selection, his primary mechanism of evolution, were less successful. The whole thrust of the theory of natural selection depends on the existence of hereditary differences among individuals, and as Darwin himself clearly recognized, the theory would rest on shaky ground until the nature of heredity was better understood. Although Darwin and Galton and many others attempted to develop theories of heredity, the beginnings of genetics as a science had to await the rediscovery of Mendel's laws in 1900. Therefore, lacking any useful knowledge of heredity, so essential to understanding the mechanism of evolution, Darwin was fumbling in the dark in his efforts to build a sound foundation under the theory of natural selection. Without an understanding of the origin, nature, and transmission of genetic variation, Darwin faced a frustrating and almost hopeless task in trying to understand how natural selection worked. His retreat into Lamarckism and his theory of pangenesis both bear witness to his difficulties.

The evidence available to Darwin was sufficient to tell him something about evolutionary history but relatively little about the mechanism of evolution. The fossil record, for example, is the best evidence for the actual course of evolution in the past. With rare exceptions, however, fossils consist not of the actual remains of individuals but of traces of their existence, most often petrified hard parts. Thus, what is available is usually a faint three-dimensional image of certain parts of an organism that lived long ago. Paleontologists have made remarkable progress in reconstructing the course of evolution from this fragmentary evidence, but they should not be expected to provide much information about the evolutionary mechanism itself. The recent efforts to promote punctuated equilibrium (Eldredge and Gould, 1972; Gould, 1982b) as a new evolutionary theory and to resurrect Goldschmidt's (1940) systemic mutations or "hopeful monsters" to explain "macroevolution" illustrate how far astray paleontologists can go when they try to deal with the process rather than the products of evolution.

The only other source of information about evolution, apart from the fossil record, is the vast array of living species. Darwin studied the geographical distribution, systematics, morphology, and embryology of living species, and drew many inferences about evolution from his studies. Although we know much more about these subjects today, they still tell us more about phylogeny than they do about the evolutionary process. Darwin studied artificial selection in domesticated species to learn more about how natural selection might work in wild populations, but again his lack of knowledge about heredity hampered his progress. Only as knowledge of genetics, cytology, and more recently, molecular biology increased has it become possible to study the genetic system itself and thus the mechanism of evolution.

The Genetics of Evolution

Evolutionary change was originally described by Darwin as "descent with modification," a type of evolutionary change known today as phyletic evolution or anagenesis. The other major form of evolutionary change is speciation or cladogenesis, the splitting of a single line of descent into two or more separate species. Ever since it was realized that in the absence of the factors actively promoting evolution, natural populations would remain in equilibrium, it has been possible to redefine evolution as a shift in the Castle-Hardy-Weinberg equilibrium. Furthermore, the unit processes underlying all evolutionary change are gene substitution and gene frequency change. The question then becomes how best to study evolutionary change, how and where to study the gene substitutions and changes in gene frequency that underlie all evolution.

The fossil record will hardly do, for fossil genes and chromosomes are extremely rare. Among living species, there are many choices. One approach is to study natural populations of a species; the poly-typic species that have differentiated into geographical races or into ecotypes are especially informative. Domesticated species, which have only been domesticated since the invention of agriculture some 10,000 years ago, had an understandable appeal to Darwin, for they have undergone some striking changes in a relatively brief time, evolutionarily speaking. They continue to provide a rich source of information because animal and plant breeders study them so intensively.

Both accidentally and deliberately introduced species are excellent subjects for study because they often undergo genetic change as they adapt to their new physical and biological environments. Populations living in disturbed habitats and populations exposed to pesticides, herbicides, antibiotics, chemical and physical mutagens, and other types of environmental agents will also provide useful information. Often such populations must evolve rapidly to avoid extinction, and the evolution of resistance or tolerance to changed environmental conditions is a rich source of information about the evolutionary process.

Many questions about the process can best be answered through the use of experimental laboratory populations under carefully controlled conditions. Although lab studies are sometimes dismissed by field biologists as unrealistic, they actually involve natural populations in a somewhat unnatural environment. The results are just as valid as those from other types of studies and sometimes provide answers that cannot be gained in any other way. The best approach is to use the lab studies to complement and extend the studies of populations in the field. A combination of field and laboratory work holds the greatest promise of providing insight into the workings of the evolutionary mechanism.

Gradual Evolution

Although evolutionists have had many disputes about the nature of evolutionary change, on one subject they have nearly all agreed: evolution is gradual. Darwin (1871, p. 361), for example, wrote, "As natural selection acts solely by accumulating slight, successive, favourable variations, it can produce no great or sudden modifications; it can act only by short and slow steps." Apart from a few mavericks like Goldschmidt, who postulated that systemic mutations were the basis for macroevolutionary change, a view that has been widely and justly criticized, the belief that evolution is gradual has prevailed ever since Darwin. Fisher and Wright both strongly believed in gradual evolution as have Crow and Lande more recently, to cite just a few. The prevailing view can best be summarized by a recent quotation from Thomson (1988): "Evolutionary theory, in whatever its neo-Darwinian, postclassical Darwinian, or post-New Synthetic guise, is quintessentially a theory of gradual change. It is a theory of continuity, and evolutionary change must therefore, by definition, be the result of mechanisms that only work through very small changes."

The odd thing about this is that the evolutionary theory developed by population geneticists focuses primarily on the genetic changes at a single locus, and occasionally tries to deal with the effects of two loci, but the math gets quite intractable when more than two loci are treated simultaneously. Segregation at one or two loci ordinarily gives rise to clear-cut, qualitative differences rather than the quantitative variations thought to underlie gradual evolutionary change.

A quite different approach has been developed to deal with the genetics of quantitative traits, which is usually referred to as quantitative or biometrical genetics. As Kempthorne (1977) stressed not long ago, even though on the surface, population genetics and quantitative genetics might seem to have a lot in common, in fact they use quite different approaches to the genetics of populations and "an expert in population genetics may not be an expert in quantitative genetics and vice versa." Because most of the economic traits of interest to animal and plant breeders are quantitative traits, quantitative genetics was developed primarily to aid them in their work and has always had a strong practical bent.

The most widely used book on quantitative genetics is Falconer's *Introduction to Quantitative Genetics* (1989). The format of the book bears out Kempthorne's contention about the separation between population and quantitative genetics. The first five chapters provide a clear, concise review of population genetics while the rest of the book is devoted to quantitative genetics. Even though both sections of the book are devoted to the study of the genetics of populations, they have very little in common with respect to content or methodology.

Even more remarkable is the failure to mention DNA and RNA until the next-to-last paragraph in the book. In other words, population genetic theory and quantitative genetic theory seem to have developed quite independently despite their common origins in Mendelian genetics and also seem to have been uninfluenced by the remarkable advances in molecular genetics in recent decades. In fact, despite their mathematical and statistical sophistication, both bodies of theory seem to involve rather primitive and simplistic genetics and have been virtually untouched by recent advances in the field. In this respect, quantitative genetics, with its dependence on means, variances, covariances, and heritabilities, seems somewhat more primitive than population genetics.

Surprisingly, recent trends in the study of evolution, as reflected in the journal *Evolution*, seem to show an increasing number of quantitative genetic studies of evolution, which in some ways seems like a retreat into the Dark Ages so far as gaining an understanding of how evolution occurs. The main reason for this trend is the nearly universal belief that evolution is gradual, involving quantitative traits, and that this is the only feasible approach to the study of the mechanism of evolution.

Lande has been one of the foremost recent leaders in such work, and his orientation is clear from the following passage (1981): "These results strongly support the neo-Darwinian theory that large evolutionary changes in quantitative characters usually occur through the accumulation of multiple genetic factors with relatively small effects, because mutations with major effects almost universally have deleterious pleiotropic effects. Although a multitude of major mutations are known to geneticists and to plant and animal breeders, there is little present evidence that single-gene mutations with major effects have been important in the evolution of natural populations." This quotation is reminiscent of a much earlier statement by Darlington (1956): "While marker genes, the chief legacy of classical genetics, with their pedigrees and their mutation rates, are of great importance for the study of evolution, they are of little importance in carrying it out." Lande (1983) wrote further: "There is no question that a wide variety of mutations with major effects do occur spontaneously but there is at present relatively little evidence that they serve as the basis for adaptive evolution in natural populations. As will be shown below, adaptive evolution by mutations with major effects occurs most often in domesticated or artificially disturbed populations."

The Role of Dominant Genes in Evolution

I believe, however, that Lande has misread the evidence and that on the contrary, evolution in natural populations more often involves genes with major effects than does evolution in domesticated or laboratory populations where numbers and available genetic variability are both limited. As a result, gradual quantitative evolutionary changes are more apt to be observed in domesticated or laboratory populations than in natural populations. The quotations from Darlington and Lande, however, are representative of prevailing attitudes about evolution that have persisted for decades.

The evolution of DDT resistance in insects has been studied in a number of cases in both laboratory and wild populations. Although no complete survey is available for all such studies, it is undoubtedly safe to say that the genetic mechanism underlying DDT resistance in laboratory populations involves multiple factors far more often than in wild populations (e.g., Dapkus and Merrell, 1977). In the wild populations dominant or partially dominant genes for resistance have been reported far more often than in laboratory populations (Crow, 1966). The most plausible explanation for this difference is that the wild populations are large and contain a much wider range of genetic variability than the relatively small laboratory populations. Even within the same species, different genetic mechanisms for DDT resistance have been reported.

An even more striking case of the importance of single genes in evolution is the evolution of industrial melanism in many dozens of species of moths (Kettlewell, 1973). In these species, almost without exception, the melanism results from the effects of a single dominant or semidominant gene. Both DDT resistance and industrial melanism occur in what Lande may consider to be "artificially disturbed populations," but this poses a problem in semantics. I regard humans and all their multifarious and sometimes nefarious activities as a part of the natural world and do not regard either DDT or industry as artificial or unnatural. Despite the longing by many biologists to do research in environments undisturbed by humans, such a goal seems to me to be unrealistic if not unattainable. Humans are a part of the natural world and of the environment of nearly all species, and research on the evolution of DDT resistance and industrial melanism is just as valid as research on the evolution of brachiopods or dinosaurs and may be considerably more enlightening.

I wish to suggest that genes of major effect, particularly dominant or partially dominant genes, do play a significant role in the evolution of natural populations and that studies of these genes hold considerably more promise of advancing our understanding of the evolutionary process than studies of the genetics of quantitative traits (Merrell, 1969b).

The importance of such genes to evolution seems to have been dismissed for several reasons. One is the oft-mentioned detrimental effect that major genes are apt to have. However, this is another generally accepted truism that is not always true. Studies of the fitness of a number of both dominant and recessive mutants with clear-cut phenotypic effects in *Drosophila melanogaster* (Merrell, 1953b, 1965; Merrell and Underhill, 1956a) showed that these mutants varied widely in fitness. Some were rapidly eliminated when placed in competition with their wild-type alleles in laboratory populations, but others were able to persist in the populations for many generations. Separate tests of various components of fitness showed that the most striking differences in fitness involved mating success. The mutants most rapidly eliminated from the experimental populations coincided with those with low male mating success. In viability and other fitness components, the mutants did not differ greatly from the wild type; mating success seemed to be the trait most sensitive to the adverse effects of the mutants. Where mating success was relatively good, the mutants were able to compete most successfully with their wild-type alleles. Thus, to say that genes of major effect have deleterious effects on fitness is not completely accurate. Some do, but not all.

Furthermore, the fitness of a gene is relative rather than absolute and may vary with its environmental and genetic background. Thus, a gene detrimental under one set of conditions may not be harmful if the conditions change. Most populations of *Drosophila* are not naturally resistant to DDT, for example, but exposure of a genetically variable population to DDT almost invariably leads to increased DDT resistance (Merrell and Underhill, 1956b). If the population is removed from DDT, the DDT resistance tends to be lost. The simplest interpretation of such results is that the genes for DDT resistance were present in the original population at low frequency, but the change in conditions (i.e., the addition of DDT) enhanced their fitness and their frequency increased. With the removal of DDT, selection favoring the genes for resistance is relaxed, and they tend to be replaced by other alleles with greater fitness in the DDT-free environment.

Similarly, the effects of a mutant may be considerably influenced by its genetic background as well as by the environment. The residual genotype is always a factor in the expression of a gene so that a gene harmful on one genetic background may be much less detrimental on another. Thus, to state that genes of major effect cannot play a significant role in evolution because they are generally detrimental is overly simplistic. If environmental conditions change, they may then be favored by selection. Furthermore, if they do become favored by some new selective force but retain some of their detrimental qualities, natural selection can modify the residual genotype so as to suppress the harmful side effects of the major genes.

Finally, it should be added that to speak of selection for genes for DDT resistance is merely a shorthand way of saying that flies carrying such genes are more resistant to DDT and thus survive and reproduce better than flies lacking such genes. This point bears repeating because of the current tendency to give credence to the idea that there are levels of selection or hierarchies of selection, ranging from gene to trait to individual to group to species and even to community (Brandon and Burian, 1984; Depew and Weber, 1985). The primary process in the case of DDT resistance is individual phenotypic selection, not genic selection, even though it is much simpler to speak in those terms. The genes do not live or die, the flies do. The susceptible flies stagger about, collapse, quiver, and die; resistant flies exposed to DDT will die of desiccation before they die from the effects of DDT.

Another reason for the belief that most evolution is gradual and involves quantitative traits is that most of our direct experience with evolutionary change comes from domesticated species, from laboratory populations, and from the fossil record. Most of the traits of economic interest in domesticated species are quantitative traits, and the efforts of breeders are aimed at increasing the yield of corn or milk or sugar or bacon or eggs produced by the species they work with. Similarly, with laboratory populations the selection experiments have usually been directed toward quantitative traits in mice or *Drosophila*, the two species most often used in such work. The fossil record provides a record of changes in morphological traits, and again students of the fossil record tend to deal with quantitative differences in morphological traits. Small wonder that biologists tend to think of evolution in terms of gradual changes in quantitative traits.

Industrial Melanism

The question to be addressed now is how evolution actually occurs in these cases of industrial melanism and DDT resistance. Why should dominant or partially dominant genes so often be responsible for these adaptive shifts? In the case of industrial melanism, the phenotypic change is relatively simple. The moths normally rest in the open on the lichen-covered bark of trees during the day, and their light-colored cryptic pattern blends in with the lichens so that predators seldom discern them. With the coming of the Industrial Revolution, soot from innumerable chimneys killed the lichens and darkened the tree trunks so that the light-colored moths became highly visible to predators. Within less than a century, the phenotypes of the moths in the polluted areas became dark to match the darkened tree trunks.

This rapid change was not the result of physiological adaptation by individuals, but was an evolutionary change, mediated in nearly all cases by a single dominant gene. Natural selection rapidly increased the frequency of these genes in populations in polluted areas as was demonstrated by Kettlewell (1973) and his associates, who showed that differential predation by birds eliminated the dark forms in unpolluted areas and the light forms in polluted areas. In recent years, improved pollution control has already led to some reversal of this selection pressure, and the populations have already begun to respond (Cook, Askew, and Bishop, 1970; Bishop and Cook, 1975; Cook, Mani, and Varley, 1986).

These studies suggest why the melanistic phenotype has replaced the light cryptic phenotype in polluted areas, but not why the evolutionary shift was so often mediated by dominant or partially dominant genes. To understand the role of dominants, we need to consider how a previously well-adapted population responds to a new selection pressure. When the environment changed so that the dark phenotype was better adapted to the polluted environment than the existing light-colored wild type, this new phenotype could be approximated by various genes or combinations of genes. In the case of melanism in moths, it was shown (Ford, 1953) that darker moths could be produced by a simple dominant or partially dominant gene, or by a recessive in the homozygous condition, or by the combined effects of genes at several loci (i.e., multiple factors). If all of these mechanisms for producing the melanistic phenotype are present in the population, they will all be favored by natural selection. If none is present, the adaptive shift to the new wild type cannot occur unless, or until, it becomes available by mutation, by migration, or possibly, by recombination. In large wild populations it is entirely possible that all three possibilities can coexist in the gene pool at low frequencies. If this were the case, it is instructive to consider which genetic mechanism has the greatest likelihood of becoming the genetic basis for the new wild phenotype when selection starts to favor melanism.

If the melanism is mediated by multiple factors with low frequencies, progress in the desired direction will be slow because sexual reproduction constantly tends to break up favorable combinations of genes at two or more loci (Haldane, 1932). Progress will also be slow if selection favors a rare homozygous recessive melanistic individual because virtually all such moths will perforce mate with light-colored individuals, nearly all of which will be homozygous for the light allele. Therefore, with rare exceptions, all of the offspring of a rare homozygous recessive melanistic individual will have light phenotypes rather than the favored dark phenotype, and selection will be unable to act in the next generation.

Selection for any dominant or partial dominant that produces the favored dark phenotype will be more effective than selection for any other type of genetic control, however, because half the progeny of a dark dominant heterozygote will also be dark and will be subject to favorable selection again in the next generation. Even with a low initial frequency, a favorable dominant has a reasonably good chance of becoming established in a finite breeding population. Therefore, of the various genetic means for reaching a dark adaptive phenotype, the most efficient is selection for a dominant or partially dominant mutant. If multiple factors, and dominant and recessive genes for melanism, are all present in a population at low frequency, selection will favor all the types of genes for melanism, but the ones that become established in the population will almost invariably be the dominants because of the greater efficiency of selection in their favor (Haldane, 1932; James, 1965). If no dominant allele is available in the population, other genetic mechanisms may become established.

DDT Resistance

The evolution of industrial melanism involves just one type of phenotypic change, from a lighter to a darker moth. The evolution of DDT resistance is more complex because resistance can be mediated by a variety of phenotypic changes. For example, when the walls of houses were sprayed to control malaria, the mosquitoes landing on walls were killed and a behavioral change evolved in the surviving population, which no longer tended to land on walls. In other cases, the cuticle of resistant insects was less permeable to DDT than that of their susceptible relatives; or the fat bodies, which accumulate fat-soluble DDT, were larger in resistant insects; or the nerves were found to be less sensitive to DDT; or an enzyme that breaks DDT down to the relatively harmless DDE was present in increased quantities. Sometimes different mechanisms of resistance were found in different populations of the same species. Such a finding should not be surprising. In the presence of a lethal agent like DDT, any mechanism, from behavioral to biochemical, that enhances the insects' chances of survival will be favored by natural selection, and the genes responsible will increase in frequency. Evolution under these circumstances has to be opportunistic; any mechanism available will be seized on by natural selection. If favorable dominants are present in the population, they will probably provide the underlying genetic mechanism, but if not, other genetic mechanisms will be pressed into service. It is small wonder that such a variety of modes of inheritance and mechanisms of resistance have been reported in insect populations.

The foregoing discussion postulated natural selection of rare dominant mutations in wild populations as the primary instrument for effecting the adaptive shifts to melanism and to DDT resistance. This postulate raises questions about the origin of these mutants, about the origin of dominance, and about the origin of the dominance of the wild type. The answer to the first question is relatively simple. The origin of the genes for melanism and for DDT resistance, like the origin of all genes in populations, can ultimately be traced back to mutation. The origin of the dominance of these genes is another matter, however.

The Origin of Dominance

A word now on the nature of dominance. Dominance is one of the first concepts students of genetics encounter, but has more intricacies than may at first be apparent. Mendel coined the term after he observed that the hybrid offspring of parents homozygous for different alleles at the same locus were not intermediate in appearance but tended to resemble one or the other parent. The trait that appeared in the hybrid or heterozygote was called dominant; the one not expressed was called recessive. The alleles themselves are sometimes called dominant or recessive, but such usage requires caution. When an allele has pleiotropic effects, it may act as a dominant on one trait but not on another. From a cross between round and wrinkled garden peas, for instance, the peas are round in the F_1 hybrids, but their starch grains are intermediate in size and shape. Thus, the allele for round is dominant in its effect on pea shape but not in its effect on the starch grains. Hence, dominance is not a fixed property of an allele. Even for a given phenotypic trait, the dominance relations are not constant. The external environmental conditions (e.g., temperature), the internal environment (e.g., hormones), and the rest of the genotype (through allelic and epistatic interactions) may influence the expression of dominance.

Even though there are no simple laws of dominance comparable to Mendel's laws of segregation and independent assortment, certain generalizations are possible. If two or more alleles occur at a locus, one is usually dominant to the others. Moreover, the dominant allele ordinarily has a higher frequency than the others. Because the wild-type allele is best defined as the most frequent allele in a population, this means that wild-type alleles are usually dominant. Thus, we are really confronted by two questions. First, what is the origin of dominance; that is, why is one allele dominant over another? And second, why are the wild-type alleles in natural populations so often dominant?

These two questions are usually regarded as parts of the same problem, but a single example can suffice to show this to be questionable. Polydactyly (excess digits) is a rare trait in humans, inherited as a simple dominant to pentadactyly (five digits per appendage), the far more common homozygous recessive condition. In this case, the recessive allele for pentadactyly is clearly the wild-type allele. Thus, even if a satisfactory theory can be developed to account for the usual dominance of the wild type, it will fail to explain the dominance of traits like polydactyly. For this reason, the origin of dominance and the origin of the dominance of the wild type must be treated separately.

Although a number of theories concerning dominance have been proposed, none has been completely satisfactory. We shall try to incorporate the best features of each into a single concept that will account for the phenomena observed. The most enduring and bitter controversy in the history of evolutionary biology arose between R. A. Fisher and Sewall Wright over the origin of dominance. Fisher (1928a, b, 1931) postulated that dominance of the wild type had to evolve, that dominance was gradually built up by natural selection for modifying factors in heterozygotes. He assumed that when a harmful mutation first occurs in a species, dominance will be lacking, and the heterozygote will be intermediate in expression between the homozygotes and therefore less fit than the existing homozygous wild type. Any genes at other loci tending to make the heterozygotes' phenotype more like that of the homozygous wild type will be favored by natural selection. In time, selection will build up a system of modifying factors that will make the wild-type allele dominant to such a recurring deleterious mutant.

One oddity about Fisher's theory is that he violated one of the rules of logic when he assumed that mutations, when they first occur, will lack dominance. This assumption is untestable and even worse, deals with the very point for which proof is being sought. A new mutation will be expressed within the framework of the existing genetic system. Its effect in the heterozygous condition will depend on its ability to function in relation to its allele, to the rest of the genotype, and to the environment. Therefore, it might show any degree of dominance initially even though the degree of dominance may subsequently be modified owing to changes in the gene complex or in the environment. There is, however, no *a priori* reason to suppose that new mutations will always be intermediate in expression in heterozygotes.

Wright's criticism (1929a, b) of Fisher's theory of the origin of dominance precipitated the bitter controversy that lasted for decades. Wright emphasized two major points. First, the heterozygotes for rare deleterious mutants will be so infrequent that selection for dominance modifiers in the heterozygotes will be far too weak to be effective. Second, these modifying factors will have their own primary effects on the phenotype, and their frequency in the population will depend on the nature of the selection pressure for these primary effects rather than for their effects on dominance at another locus.

As an alternative to Fisher's theory, Wright (1934) proposed a "physiological" theory of dominance, in essence arguing that dominance depended on the inherent properties of the gene as expressed in a particular genetic and environmental background. His theory postulated that dominant alleles are functional while recessive alleles are either partially or completely inactive. The levels of dominance were thus related to the levels of enzymatic activity of the alleles. This theory was a more sophisticated version of the discredited earlier "presence-absence" hypothesis of Bateson and Punnett. Wright supposed that for most traits there is an upper limit to the expression of the trait, and that if one dose of a given allele suffices to reach that limit, then the allele will act as a dominant because two doses will have no more effect than a single dose, and the heterozygotes and homozygotes will have similar phenotypes. Wright's theory seemed to imply that dominance was related to the inherent properties of the dominant allele. Muller's (1932b) studies of hypomorphs and amorphs, which represented partial or complete inactivation of the wild-type alleles, seemed to support Wright's concept. However, Muller also described hypermorphs, antimorphs, and neomorphs, which were rare alleles usually dominant to the wild type. In this sort of framework, polydactyly would probably be categorized as a hypermorph. Furthermore, in some cases each allele is expressed independently, and neither is dominant to the other. Individuals with the AB blood type produce both antigen A and antigen B in their red cells, and the alleles are said to be codominant. These and similar results show that different alleles may produce qualitative as well as quantitative differences and that alleles may act quite independently of one another.

Wright's physiological theory can account for dominance of the functional over the nonfunctional allele and also for dominance of the wild-type allele because a functional gene is ordinarily needed to produce a "normal" or wild-type phenotype. However, neither Wright's nor Fisher's theory can adequately explain all of the phenomena associated with dominance. Fisher's modifier theory fails to account for the dominance of such rare dominant traits as polydactyly or for the presence of rare dominant alleles for melanism in so many species of moths. In fact, as we shall soon see, adherence to Fisher's modifier theory led to some very tortuous reasoning in relation to the origin of industrial melanism. Wright's theory, on the other hand, with its emphasis on gene action, fails to account for the undoubted modifiability of dominance.

Haldane (1930, 1939) proposed a somewhat different version of Wright's theory to explain the origin of dominance of the wild type. His theory involved the selection of more efficient wild-type alleles. He suggested that a number of isoalleles may exist, all of which approximate the wild-type phenotype when homozygous, but vary in expression when heterozygous with a deleterious allele. The more efficient isoalleles, that is, those that produce the closest approximation to the well-adapted wild-type phenotype when heterozygous, will be favored by natural selection and will tend to replace the less efficient alleles in the population. The demonstration of isoalleles by Stern and Schaeffer (1943) made this theory both plausible and appealing in its simplicity.

H. J. Muller (1932b) and C. R. Plunkett (1932, 1933) independently proposed a theory for the origin of dominance of the wild type through the selection of modifiers that was different from Fisher's. They had observed that the wild-type phenotype when subjected to environmental stress during development is more stable than mutant phenotypes; that is, it has greater developmental homeostasis. This developmental homeostasis, they suggested, results from the selection of modifiers that buffer the developing wild-type embryo against the harmful effects of environmental stress. The selection for these modifiers occurs in the frequent wild-type homozygotes rather than only in rare heterozygotes for a deleterious mutant, as Fisher proposed, and thus met one of Wright's major objections to Fisher's theory.

Support for this version of the modifier theory comes from Rendel's work with scute (1959, 1962), which demonstrated that, owing to the effects of modifying factors, the wild type is phenotypically more stable than mutant phenotypes. Under this theory, the dominance observed in heterozygotes is an incidental by-product of the canalizing selection for stability of the wild phenotype.

It is worth pointing out that the modifier theories and the gene action theories of dominance are not mutually exclusive but complementary. Our problem is to sort out the reasons for the origin of dominance and for the dominance of the wild type. Genes such as polydactyly are dominant despite the fact that they are not wild-type alleles, and some version of Wright's theory seems best able to account for dominance in these cases. The origin of the dominance of wild-type alleles, however, is of greater significance and poses more complex, but also more interesting, problems.

The Origin of Dominance of the Wild Type

The simplest means of origin for dominance of the wild type is for an existing wild-type allele to be replaced by a rare dominant mutant that generates better-adapted phenotypes than the existing allele. Apparently this is what happened with industrial melanism, for Ford (1975, p. 333) wrote, "The fact that nearly all industrial melanics are complete dominants when they begin to spread poses a considerable problem." The reason Ford felt it posed a problem was that he was a strong advocate of Fisher's modifier theory and expected the dominance of the melanism genes to evolve gradually rather than to be present from the start. To circumvent this supposed problem, Kettlewell (1973) suggested that the dominance of the genes for melanism had evolved during the Boreal period some 10,000 years ago when most of Great Britain was covered by pine forests. Present-day relict pine forests contain melanic forms, suggesting that melanism could have been favored during the Boreal period and the dominance modifiers accumulated at that time. Then when the Industrial Revolution came, the gene pool of these moth species already contained the modifiers needed to make the alleles for melanism dominant. However, this argument has a serious flaw. If modifiers making the melanic alleles dominant had accumulated during the Boreal period, then during the subsequent thousands of years when the light-colored cryptic form was the favored wild type, those modifiers should have been supplanted by others that enhanced the dominance of the light-colored cryptic form. Thus, this *ad hoc* explanation seems rather forced and stems from an attempt to reconcile the facts with Fisher's theory.

Another aspect of the evolution of melanism does not bear up under scrutiny. Ford (1937, 1940) proposed that the melanic forms were hardier or more viable than the light-colored cryptic types but never became established because their selective advantage in viability was offset by their lack of protective coloration in the unpolluted country-side. A reexamination of the data on which this supposition was based (Merrell, 1969b, 1981) failed to support this conclusion, how-ever. The data came from single-pair matings in a number of different moth species. In the great majority of crosses, the expected 1:1 or 3:1 ratio of melanic to typical forms was observed, but in a few excep-tional cases there was an apparent excess of the melanic form. A breakdown of the data in these exceptional cases showed that the ex-cess was traceable to just one or a few broods, with all the others showing the expected 1:1 ratio. In these exceptional broods there was a significant deficiency of the typical form, suggesting that these ho-mozygous typicals ($+/+$) were also homozygous for linked deleteri-ous or even semilethal genes. However, the heterozygous (M/$+$) mel-anics from these same matings, with chromosomes of different origin, were unlikely to be homozygous for harmful alleles. Lumping these heterogeneous broods together gave the spurious appearance of somewhat greater viability for the melanic type compared to the light typical form. More recently, Cook, Mani, and Varley (1986) have also concluded that the United Kingdom data on melanic morph fre-quency in the peppered moth, *Biston betularia*, can be adequately ex-plained without the assumption of heterozygote advantage.

Under the concepts set forth here, it should not be surprising that the rapid transition to melanism in industrial areas was mediated by a dominant in so many different species. If suitable dominants are present in the gene pool of a population, they have a high probability of being seized on by natural selection to effect the change to a better-adapted phenotype. In some parts of the range of the peppered moth, melanism is mediated by a completely dominant gene known as *car-bonaria*. In other parts of the range, a partially dominant melanic allele known as *insularia* has become established. Presumably, this repre-sents an example of the opportunism of evolution. In the absence of the *carbonaria* allele, selection seized on *insularia* as the next best thing available. If selection for melanism continues, the fate of *insularia* would be interesting to follow. It might be replaced by *carbonaria* through migration and gene flow; it might be supplanted by selection of a more efficient allele arising through mutation *a la* Haldane; or its dominance might be enhanced through the accumulation of modifiers if neither of the other possibilities transpired. However, the control of atmospheric pollution appears to have reduced the selective advan-tage of melanism and has led to a decrease in *carbonaria* frequency (Cook, Mani, and Varley, 1986), an impressive display of the rapidity with which natural selection can track changing environmental conditions.

In many cases of melanism and insecticide resistance, the newly favored alleles appear to have been dominant from the outset. There appears to have been no buildup of dominance through the selection of modifiers. However, when such dominants are crossed into populations other than the ones in which they naturally occur, the dominance sometimes breaks down (e.g., Sheppard and Ford, 1966). This result has been interpreted to mean that the dominance of these genes must have evolved in their native gene complexes through the selection of modifiers enhancing dominance. The breakdown of dominance on outcrossing by no means proves this theory, however; all it shows is that a gene dominant in one genetic background may not be dominant in another. Only direct observations and experiments during the period when a newly favored allele is becoming established in a population can determine its dominance relations at that time. This is not to say that dominance can never be enhanced by modifiers, but rather that the breakdown of a gene's dominance on outcrossing does not prove that it originated through the selection of modifiers in its native population.

The major difference between the ideas outlined earlier and previous theories of the origin of dominance of the wild type is one of perspective. Rather than focusing on a single locus, we have focused on the phenotype, asking what happens when a previously well-adapted population is subjected to a new selection pressure. With this broader perspective, it is clear that dominance can be achieved in a variety of ways, that the dominance of the wild type is a means to an end—namely, a well-adapted phenotype—rather than an end in itself. When strong new selection pressures are applied to a population and the best-adapted phenotype suddenly changes, the population must adapt quickly or become extinct. Since rapid adaptive change is mediated more effectively by dominant genes than by any other genetic mechanism, dominants must play a more significant role in evolution than has been generally recognized. The idea that evolution is gradual, with very small constant selection coefficients of the order of 0.01 or 0.001, is simply not consistent with the natural history of any species (Endler, 1986, chapter 7). Its members are always facing new competitors, new predators, new parasites, new diseases, and ever-changing physical conditions. Even though, on a geological time scale, evolution may appear to be a slow, plodding process if measured by the increase in size of a tooth cusp, that is hardly the way evolution occurs if one tries to follow the vagaries of existence of a breeding population on a day-to-day basis. Then all is in a state of flux, with changing selection, mutation, and migration pressures, plus an occasional modicum of random genetic drift.

The Evolution of Mimicry

As it may be argued that the evolutionary changes involved in industrial melanism or insecticide resistance are too artificial or too simple genetically to be truly representative of the evolutionary process, let us turn our attention to mimicry, which can hardly be subject to such criticisms. Mimicry has always posed a challenge to Darwinian theory because the gradual accumulation of "slight, successive, favorable variations" seems an unlikely explanation for the evolution of mimicry.

Ever since the original papers by Bates (1862) and Müller (1878), an abundant literature has been published on mimicry and on the related subjects of cryptic and warning coloration. Those who wish to delve further into these fascinating subjects can enter the literature through such works as Punnett (1915), Nicholson (1927), Fisher (1930), Carpenter and Ford (1933), Cott (1940), Goldschmidt (1945), and more recently, Wickler (1968), Ford (1975), Sheppard (1975), and Turner (1977, 1983). It is not our purpose to deal with the varieties and complexities of the many systems of mimicry outlined in Wickler, but rather to concentrate on the genetic mechanisms that make adaptive coloration possible.

Many animals are cryptically colored, avoiding the notice of predators by blending in with their surroundings. The evolution of cryptic coloration seems rather straightforward, for any heritable variations that enable animals to escape detection by predators will permit them to survive and reproduce. The song sparrow, a polytypic species in which some thirty-four subspecies have been identified, is widely distributed over much of North America in a variety of habitats. The color of the birds subtly changes across its range, lighter in the desert areas and darker in the more humid regions, suggesting not only that the coloration is adaptive but that it is constantly being fine-tuned by natural selection. Perhaps for the skeptics it is worth adding that there is convincing experimental evidence for the protective nature of cryptic coloration, that the concept is not just the product of some biologist's overactive imagination. Cryptic coloration is a bit of a misnomer, however, for the crypticity often involves pattern, structure, and behavior as well as color.

Other animals are so conspicuously or aposematically colored that they seem to announce their presence to predators. Such species are usually unpalatable like the monarch butterfly or noxious like the skunk or armed with a sting like the hornet. This, too, is a form of adaptive coloration, for by announcing their presence in unmistakable terms, these animals warn off potential predators and thus live to play another day. The adaptive story in this case is not quite so simple as that of cryptic coloration, for a naive predator may never have gotten the message and may blunder about trying to feed on warningly colored individuals. Given the unpleasant nature of the message, however, few predators remain naive for very long. The experimental evidence on single-trial learning has shown how immediate and long-lasting is the impact on a predator of a single experience with a noxious prey organism.

For warning colors to evolve, first the noxious quality must develop. The sting of a hornet is a modified ovipositor whereas the foul-smelling musky liquid sprayed by skunks is produced by modified scent glands. Neither of these developments poses any particular challenge to natural selection or to the imagination of biologists. If an ovipositor is used successfully as a weapon or the musk from a scent gland deters a predator, the efficacy of these deterrents can quite easily be improved by natural selection. If, while these noxious qualities are evolving, the more conspicuous individuals are spared most often by predators, aposematic colors will be favored by natural selection, and the evolution of noxious qualities and warning coloration will proceed hand in hand. The monarch butterfly is apparently a case of evolutionary serendipity, for its unpalatable taste comes from the milkweed plants on which it feeds rather than from its own body chemicals, and not all populations are unpalatable. Nonetheless, warning coloration has evolved in the monarch despite the external source of its nauseating taste.

Müllerian Mimicry

At this point the story adds another level of complexity. Müller noticed that not only did dangerous or distasteful organisms advertise their presence, but whole groups of more or less distantly related species like the bees and wasps tended to resemble one another. Although this phenomenon is now called Müllerian mimicry, it is not mimicry at all in the usual sense because there is no model, no mimic, and no deception. What is involved is a numbers game.

For protection, harmful insects like bees and wasps depend on their striking yellow and black patterns to announce their presence to potential predators. If each harmful species had its own distinctive warning system, each predator would have a whole series of painful lessons to learn, and the toll on the prey species would be correspondingly great. However, if potential prey such as bees and wasps resemble each other enough to cause a predator, once stung, to hesitate before attempting to capture a member of a similar species, natural selection will tend to preserve and enhance any similarities among the species. Selection pressure then will favor convergent evolution in the bees and wasps toward a common and increasingly conspicuous pattern of warning colors. However, in Müllerian mimicry the resemblance between members of a mimicry ring or group need not be very exact. Even if a predator is able to distinguish between members of different species, they will all convey the same noxious message.

The numbers game lies in the fact that the predator benefits because in a single lesson it learns to avoid a whole array of harmful species rather than just one and is thus spared a number of painful experiences. The prey species also benefit in this numbers game because many more individuals are spared during the predators' learning process.

Batesian Mimicry

Batesian mimicry further complicates the situation, for it involves a predator, an unpalatable or noxious model, and an unrelated, palatable mimic. Unlike Müllerian mimicry where the predators and the members of the Müllerian mimicry group all benefit, only the mimic benefits in Batesian mimicry. The harmless Batesian mimic, by resembling a noxious species, escapes with its life; the predator, by being duped, loses a potential meal; the models suffer greater predation because the predators receive a mixed message when preying on both models and mimics. However, if a predator can learn to distinguish between model and mimic, it will feed on the mimics and ignore the models. Thus, for a Batesian mimic there is a high premium on deceiving the predator. Although mimicry may involve external morphology and even behavior as well as color and color pattern, the resemblances are always superficial. Of course, if a Batesian mimic ever became an exact duplicate of an existing model, it would be impossible to recognize, but the phenotype's the thing, and the evolution of Batesian mimics seems to have progressed only to the point where their external appearance can deceive potential predators, and no further.

One of the requirements for both Batesian and Müllerian mimicry is that there be predators in the neighborhood capable of learning to avoid noxious organisms. For the most part, this means that the predators will be vertebrates: fishes, amphibia, reptiles, birds, and mammals, which are capable of learning and capable of being deceived. In Batesian mimicry the model is conspicuous and unpalatable or otherwise noxious to predators. At some point in their lives, the predators, the models, and the mimics must all inhabit the same place at the same time. A Batesian mimic outside the range of its model would gain no protection from its mimicry, for the predators would never learn to avoid the noxious model. Because Batesian mimics are protected from predators only by their similarity to a harmful species, their resemblance to their models will be much more carefully monitored by natural selection than the similarities between Müllerian mimics. This conclusion is supported by the fact that when a model species consists of a number of subspecies, the mimic populations change geographically to match the local subspecies of the model.

Again a numbers game is involved. If the chance that a predator will encounter the models and mimics is proportional to their numbers, then the greater the number of mimics, the more likely a naive predator will encounter them first and learn the wrong lesson—that here is a tasty morsel. Thus, a delicate numerical balance develops, determined by the numbers of the mimic, the numbers of the model, the aversive qualities of the model, and the sensibilities, perceptiveness, and hunger level of the predators. Although some efforts have been made to establish limits for the proportions of models and mimics in Batesian mimicry, such figures are not very reliable. They are so dependent on the degree of unpalatability of the models, the level of hunger and the powers of discrimination of the predators, and the precision of mimicry in the mimics that quantification is extremely difficult. If certain simplifying assumptions are made, it can be inferred that models will generally be more numerous than mimics, but even this conclusion may not hold for a particularly noxious model or for a model that appears earlier in the season than the mimic.

The palatability of prey, the discriminatory powers and degree of hunger of predators, and the deceptiveness of mimics are all continuously variable traits. In other words, some predators are more perceptive than others, some mimics are more deceptive than others, and the prey are not just edible or inedible but may fall anywhere along a continuum of palatability. The differences are relative rather than absolute. What one predator finds palatable, another may shun. In some cases a Batesian mimic of one species may act as a Müllerian mimic with another. Thus it may not always be possible to decide by simple inspection whether a particular assemblage of species represents Müllerian mimicry, Batesian mimicry, or both. Sir Walter Scott's saying "Oh, what a tangled web we weave when first we practise to deceive" (*Marmion*, 1808) seems entirely appropriate. Nonetheless, even though the two types of mimicry may not always be easy to distinguish, mimicry of both types is a widespread phenomenon in the biological world, one that demands explanation. Furthermore, the evolutionary changes associated with mimicry are not simple adaptive responses to changing physical conditions but rather are influenced or even determined by the behavior of other organisms.

We have already noted that the evolution of cryptic coloration and of Müllerian mimicry seems to pose no particular problems for modern evolutionary theory. The evolution of Batesian mimicry, however, is another matter, for it has been the subject of debate and controversy. The problem, of course, is that it is difficult to envision how a palatable species can evolve to resemble an unpalatable, unrelated, and morphologically different species by a series of slight, successive, favorable variations. Batesian mimicry has therefore become a testing ground for theories of gradual evolution, on the one hand, and theories of saltational evolution, on the other, but neither type of theory provides a completely satisfactory explanation for the origin of Batesian mimicry. Fortunately, some of the more recent experimental data on the genetics and behavior of prey and predators have begun to shed light on the question. One problem in the past has been that scientists often had such a strong commitment either to gradual evolution or to saltational evolution that other possibilities were dismissed.

The Evolution of Batesian Mimicry

The difficulty with the gradual evolution of Batesian mimicry is that it fails to explain how a potential mimic, different in appearance from a noxious species, can get started on the road to mimicry if it must perforce start out with only a very small step. If a potential mimic differs only very slightly from the other members of its own species, it can hardly show much similarity to the model and will gain no protection from predators. Yet from the very outset, the potential mimic must resemble the model enough to deceive at least an occasional predator or it will not be preserved by natural selection. It seems highly unlikely that this initial hump can be crossed by the accumulation of a series of "slight, successive, favorable variations." To circumvent this difficulty, authors from Darwin to Fisher have suggested some ingenious and convoluted explanations for the gradual evolution of Batesian mimicry.

Because of the inherent problems of the gradual theory, the saltationists proposed to get over the initial hump in a single step by a major mutation. Punnett (1915), for example, felt that the origin of Batesian mimicry could be explained by the occurrence of homologous mutations in the same genes of the model and its Batesian mimic. Years later, Goldschmidt (1945) proposed that parallel mutations were responsible for Batesian mimicry, with model and mimic using not necessarily homologous genes but rather homologous developmental pathways.

Punnett's concept was soon shot down when it was pointed out that homologous genes could hardly be involved when the same color in the same location in model and mimic was produced by two chemically different pigments, that marks on the body of a model were matched by patches of similar color on the mimic's wings, and that mimicry sometimes involved entirely different taxonomic groups, which could hardly be expected to have homologous genes. Goldschmidt's somewhat broader concept was open to similar criticisms, and both theories stumbled over the fact that mimicry often involved behavior and external morphology as well as color and pattern. It seemed highly improbable that such an array of complex changes affecting such diverse traits could be produced in the mimic by a single mutation.

In recent years it has been demonstrated that birds nesting in colonies are capable of recognizing their mates and their young as individuals amid the multitude of other very similar birds in the colony. Obviously, they are capable of detecting very subtle individual differences. On the other hand, Lack (1943) showed that a territorial male English robin would attack a bundle of red feathers much more aggressively than he would a mounted robin complete in every detail except for the red breast. This result stands in stark contrast to the colonial birds' powers of discernment and suggests that if predators are stimulated to attack prey by similar simple cues, an essential clue to the origin of Batesian mimicry may be at hand.

Another type of ethological study showed how effective single-trial learning can be in protecting mimics, once a predator has had a single unpleasant experience with a model. After two naive blue jays had each eaten a single emetic monarch butterfly, they were each offered 120 monarchs over a two-day period (Brower, Pough, and Meck,

1970). The first rejected all 120 butterflies on sight, the other rejected the first 96 before finally attacking the 97th.

These experiments suggest that the most reasonable hypothesis for the origin of Batesian mimicry is one that combines elements of both the saltationist and the gradualist theories. If a mutant in a harmless species causes it to resemble a noxious species enough to cause a potential predator to avoid it, the mutant will tend to increase in frequency in the population. The experiment with the male robin suggests that the similarity need not be exact, but neither can it be trivial or subliminal. It must represent a step in the direction of the model great enough to lead at least some predators astray. If a male robin can be misled into thinking a bundle of red feathers is a competing male robin, an obtuse predator could be deceived by a mutant that gave a potential mimic some semblance of a resemblance to a noxious model. The mimicry need not be perfect nor must every predator be deceived. All that is needed is that some selective advantage accrues to individuals showing the mutant, so that they manage to survive and reproduce better than individuals lacking it. Given what has been said earlier about the evolutionary role of dominant genes, it should be clear that the initial, crucial step toward mimicry would be expected often to involve dominant or partially dominant genes (Merrell, 1969b, 1981). Once this first step toward Batesian mimicry has been made successfully, the hump has been crossed, and selection pressures subsequently will be directed toward improving the accuracy of the mimicry.

A Batesian mimic's image may be further improved in several ways. If the initial step involves color, subsequent changes may involve color pattern, structure, and behavior as well as color. If suitable favorable mutants are available affecting pattern, structure, or behavior, they may be pressed into service by natural selection, but even if such mutants are not available, further refinements may occur through the gradual selective accumulation of polygenes or modifying factors. In this case, as in the origin of pesticide resistance, evolutionary opportunism will be at work. The initial step toward Batesian mimicry depends on the availability of a suitable mutant in the population of potential mimics. The superficiality of the resemblance between model and mimic also suggests that opportunism is at work, that any variant, large or small, that contributes to the deception of a predator will be utilized. In other words, selection operates, not to produce an exact replica of the model, but by deceiving predators with whatever means may be at hand. If one pigment is not available in the mimic, another of the same color will do. What at first glance may appear to be structural similarities in model and mimic may in reality involve different structures. The impression that emerges from the study of mimicry is not one of an all-powerful natural selection, able to accomplish any feat of adaptive legerdemain, but rather of a natural selection forced to work within the limits imposed by the available genetic variability. Thus mimicry is achieved by a striking combination of chance and adaptive events and is representative of the way all evolution occurs.

As in his theory of the origin of dominance of the wild type, Fisher's treatment of the origin of Batesian mimicry (1930) was shaped by his commitment to the concept of gradual evolution by the slow accumulation of a series of genes of small effect. Because of this commitment and because of the difficulties with the gradual evolution of Batesian mimicry discussed earlier, Fisher engaged in some first-class mental gymnastics, giving a genetical twist to an explanation first proposed by Darwin half a century before, that originally both model and mimic were similarly dull in color and that as the noxious model gradually became more conspicuous, the mimic tracked the evolution of the model. In essence, because of Fisher's belief in the theory of gradual evolution through the accumulation of modifiers, he seemed to interpret the facts to fit the theory rather than modifying the theory to fit the facts. Ford also adhered to Fisher's gradualism (1953, 1975; Carpenter and Ford, 1933), although his writing, like Fisher's, has ambiguities that make it difficult to know just what he did believe about the origin of Batesian mimicry.

The Two-Stage Theory

The two-stage theory, a synthesis of the gradualist and saltationist views, provides the most reasonable explanation for the evolution of Batesian mimicry. It has a long history, traceable back to Poulton (1912) and Nicholson (1927). After a period of eclipse owing to the influence of Fisher and Ford, the two-stage theory was resurrected, primarily by Sheppard (1961, 1975) and Turner (1977, 1983), because it provided the most plausible explanation for the results from their extensive experiments on mimicry in butterflies. In essence, the theory holds that the first stage involves a mutant of sufficient effect to deceive a predator from the outset and thus bridge the "unbridgeable gap" between model and mimic on the road to Batesian mimicry. The second stage is the establishment of genes at other loci (modifiers, polygenes, or even other major genes) that refine and polish the deceptive image of the mimic.

It seems highly probable that the crucial first step toward Batesian mimicry, like the evolution of industrial melanism and pesticide resistance, will often be mediated by a dominant or partially dominant mutant (Merrell, 1969b, 1981). In fact, a mutant of significant effect at the outset is far more crucial to the evolution of Batesian mimicry than to the origin of either industrial melanism or pesticide resistance. These mutants, however, should not be equated with the homologous mutations of Punnett or the parallel mutations of Goldschmidt, and they certainly should not be regarded as systemic mutations or macromutations. They are selected, not because they affect homologous genes or similar developmental pathways, but because they deceive predators. Again, the phenotype's the thing.

In some species, the story of Batesian mimicry is more complicated than has been indicated thus far. For example, in some cases only the females are Batesian mimics while the males are nonmimetic and presumably close to the ancestral phenotype. Moreover, the females may be phenotypically polymorphic, mimicking several different noxious butterflies in the same area, though they may also include a nonmimetic morph like the males. Finally, these polymorphic differences in the females segregate as if they were controlled by alleles at a single locus. The interpretation of such results has been a challenge, and we can only suggest some of the ideas that have been advanced. Some of the explanations seem like *ad hoc* special pleading whereas others are supported by experimental data.

The absence of mimicry in the males has been attributed to the fact that female butterflies react to visual stimuli during courtship and that males with aberrant color patterns will be less successful in mating than typical males. The mimicry of several different models by the females is accounted for by the ability of such a system to support larger numbers of the mimic species than would be possible if they mimicked just one species. The numbers game again.

The control of the polymorphic differences among the females by alleles at a single locus poses the biggest conundrum of all. Despite our dismissal of the mutation theories of Punnett and Goldschmidt, we are confronted by what appears to be strong support for single-gene theories. The evidence suggests otherwise, however. Rather than originating by a single mutation, the controlling, segregating gene locus appears to be a gene complex or complex locus or cluster of closely linked genes, which is sometimes referred to as a supergene. This supergene is apparently built up through the aggregation of a number of unlinked or loosely linked genes affecting color, pattern, structure, and behavior into a complex of tightly linked genes in which crossing over is virtually absent. The supergene acts as a switch mechanism that guides the development of the mimic into one or another of the paths that lead to mimicry of one of the models.

Although the concept that the genes governing mimicry will be assembled by natural selection from throughout the genome into a tightly linked gene complex that serves as a switch mechanism might seem rather far-fetched, experimental evidence from the work of Sheppard (1975 and earlier) and Turner (1977, 1983) provides support for the idea. Exactly how the gene complex is assembled is still in doubt. One possibility is that chromosomal rearrangements bringing the genes involved in mimicry closer together will be favored by natural selection. Another is that the supergene can only be built up by a sievelike selective process at loci where the genes to be involved in mimicry are already rather closely linked (Charlesworth and Charlesworth, 1976). This latter possibility seems unduly restrictive, given the number and variety of Batesian mimics, but until more is known about the ease and flexibility with which changes in the genetic architecture can occur, it is not possible to choose between these alternatives.

The message to be gained from this discussion of evolutionary dynamics is that single genes can and do make a difference in evolution. Despite the opinions of Darwin, Fisher, Darlington, Lande, and many others, evolution is not always a gradual process nor is it restricted to evolutionary changes in quantitative traits, resulting from the slow accumulation of numerous genes with small additive effects. Many evolutionary changes undoubtedly occur in this way, but the evidence from numerous studies of evolution in action suggests that we need to broaden the scope of our inquiries to include studies of the role of individual genes and their phenotypic effects on the course of evolution before we can hope to have a complete understanding of the mechanism of evolution.

Chapter 9
Patterns and Modes of Evolution

Thus far we have considered what might be called the unit operations of evolution, the ways in which gene frequency changes and gene substitutions occur in a breeding population. Now we need to consider evolution from a somewhat broader perspective, that is, the evolutionary changes that occur within the confines of a single species rather than within a single breeding population. Intraspecific evolutionary divergence has sometimes been called microevolution, with the added corollary that microevolutionary processes cannot account for the origin of species or higher taxonomic groups. Before we can decide whether this corollary is valid, it is essential to discuss the nature and origin of intraspecific evolutionary change.

Polytypic Variation

The patterns of variation within a species can be thought of as being of two kinds, polymorphic and polytypic. The genetic variation within a single breeding population is called polymorphic and was discussed in chapter 5. If a species consisted of a single large breeding population, all of the genetic variation within the species would be polymorphic. However, the total population of a species is ordinarily dispersed over a wide geographic area and consists of a number of more or less distinct breeding populations. When these different populations are compared with one another, they frequently show consistent differences, some of which persist even when the populations are reared in a common environment. Hence, some of the differences must be hereditary. The genetic variation among different breeding populations of the same species is called polytypic variation, and a species consisting of a number of such genetically distinguishable

110

breeding populations is known as a polytypic species. Such findings indicate that microevolutionary changes must occur within species, that a species is not a static fixed entity but rather a dynamic unit constantly subject to change. (For more on population structure, see Wright [1969], chapter 12, and Merrell [1981], chapter 11.)

Members of the same deme or breeding population ordinarily share not only a common gene pool but a common environment. The extent to which members of different demes share a common gene pool is dependent on the amount of gene flow between them. In many cases the gene flow may be minimal because, even though they are not reproductively isolated, they may be physically separated by geographical barriers, or, even in the absence of such barriers, they may be separated by such distances that individuals from the different populations never mate. Gene flow tends to keep different populations of the same species from diverging, but the other evolutionary factors all tend to favor genetic divergence.

For example, different populations of a species are finite in size, and if they occasionally go through a bottleneck in numbers, random genetic drift may come into play. By its very nature, drift is unlikely to produce similar changes in gene frequency or fixation of the same alleles in different populations, so that drift, if it occurs, will tend to produce random genetic divergence among the populations. Moreover, with low mutation rates, the arrays of mutations arising in different finite populations are unlikely to be the same, and mutation will also tend to produce random genetic differences among different demes of the same species.

Whereas members of the same deme share a common environment, it is almost axiomatic that members of different demes, since they occupy distinct areas, are confronted by physical and biological environments that vary to a greater or lesser degree. To the extent that their environments diverge, different demes will be subjected to dissimilar selection pressures and consequently will tend to develop adaptive differences from one another. Therefore, selection, mutation, and drift all tend to produce genetic divergence among separate populations of the same species. It is little wonder that most carefully studied species have been found to be polytypic. The differences among such populations can be expected, however, to manifest a combination of adaptive and nonadaptive features.

One of the questions that haunts evolutionary biologists is to what extent the differences among demes in a polytypic species are adaptive and to what extent they are merely the result of chance. Although some biologists have taken the position that intraspecific polytypic genetic variation has no adaptive significance while others have argued that such variation is primarily adaptive, these positions seem extreme, for, as indicated earlier, both adaptive and nonadaptive changes are possible. It seems unwise to attribute polytypic genetic variation to either random or adaptive events without some evidence one way or the other. However, the question cannot be resolved by simple inspection of members of different demes. Because they have not lived in the same environment, they may show environmental as well as genetic differences. Therefore, to tease apart the environmental from the hereditary differences and the random from the adaptive genetic differences in members of different demes is no simple task.

Although a plethora of names has been coined to categorize the products of intraspecific evolution, most of them are of limited usefulness because the units they purport to describe are usually not clear-cut or easily differentiated. Nevertheless, we shall discuss such infraspecific categories as ecotype, cline, ecological race, geographical race, and subspecies because of the insight they provide into the nature of polytypic variation.

If examined closely enough, differences in genetic composition could probably be found between any two demes of the same species. This interdemic variation has its origin in the chance events related to mutation and drift and in the adaptive changes brought about by selection in different environments and can persist only if there is sufficient isolation between the populations to keep gene flow from swamping their differences.

Within a species, several levels of genetic variability can be recognized. The first is the polymorphic variation within a single breeding population. The second is the variability between isolated local populations living under similar environmental conditions. Although subjected to similar selection pressures, such populations may still differ genetically, primarily as the result of mutation or drift. The third level involves differences between populations living under different environmental conditions. In this case, the effects of selection are added to those of mutation and drift.

The early studies of intraspecific variation focused on the morphological differences between geographically distinct populations of the same species. Where clear-cut differences could be seen, as, for example, in the thirty-four named subspecies of the song sparrow or between island populations in an archipelago, subspecies could be easily identified and were dignified with a trinomial name. However, not all intraspecific variation falls into such neat patterns.

Although subspecific names are usually given to geographical races and hence are based on differences in morphology and distribution, geographical races not only occupy different areas but also live in different environments so that these races must also vary ecologically to some extent. When Turesson (1922a) discovered that common widespread species of plants consisted of a number of subunits, each genetically adapted to a particular type of habitat, he called these subunits *ecotypes.* Turesson made distinctions not on the basis of distribution and morphology but on the basis of genetic adaptation to different habitats. Thus, geographical races and ecotypes merely represent different ways of viewing the same biological entities, the natural populations that comprise the species, and are as inextricably intertwined as morphology and function in an individual organism. It is worth noting, however, that similar ecotypes, adapted to life in similar habitats, may nonetheless vary morphologically from one another as the result of the random effects of mutation or drift, whereas different ecotypes may sometimes look very similar because the adaptive shifts are physiological rather than morphological (Clausen and Hiesey, 1958). The seeming discreteness of infraspecific entities may be more dependent on the sampling procedures used than on any other factor, for sampling along a transect of a species often reveals the existence of gradual continuous character gradients rather than distinct, separate populations. Both genetic and phenotypic geographic gradients, known as *clines,* have been observed.

Thus, intraspecific polytypic variation may change gradually and continuously across the range of a species or may show discontinuities, as, for example, in populations living on oceanic islands. These patterns of variation represent the extremes, and the actual pattern of variation may fall anywhere along a continuum between these extremes, depending on the population structure and the amount of gene flow among the different breeding populations. Because sampling is often sketchy, because gene flow can rapidly change the genetic characteristics of breeding populations, and because the patterns of intraspecific variation are so ephemeral, formal subspecific trinomial names should be used sparingly, primarily as a matter of convenience, and should not be taken too seriously. Too often, further sampling and more careful study of a species reveal that the subspecific names are unwarranted because the species does not consist of well-defined, discrete subunits. Too often, subspecific names must be tossed into synonymy where they clutter the literature forevermore.

As we have just seen, the factors that bring about microevolution-ary changes within a species are the same as those that lead to genetic changes within a breeding population. Mutation is not only respon-sible for the origin of variation within a deme, but may also lead to genetic divergence among demes, a role also played by random ge-netic drift. Because different breeding populations, occupying distinct areas, are isolated to a greater or lesser degree and exposed to some-what different ecological conditions, they are inevitably subjected to somewhat different selection pressures, and hence selection, as well as mutation and drift, contributes to the microevolutionary changes within a species. The counterbalancing force that tends to prevent di-vergence and ties the various breeding populations together into a single species is gene flow. Without gene flow or at least the contin-ued possibility of gene flow, each deme would form its own self-con-tained, independent evolutionary unit. Without some degree of iso-lation among the various breeding populations, the genetic divergence of these populations into distinctive infraspecific groups would not occur. Thus, the origin of races, ecotypes, clines, and the other infraspecific categories is brought about by a continuation and extension of the effects of the same factors that lead to evolutionary changes within a single population: mutation, selection, and drift.

The Nature of Species

Now we shall consider the nature and origin of species. A number of definitions of species have been proposed over the years. The most meaningful are based on reproductive isolation as the major criterion for separating species. These definitions are sometimes rather awk-wardly called "biological" species definitions in contrast to the "mor-phological" species definitions, which rely on morphological traits to distinguish one species from another.

One type of biological species definition identifies species in a neg-ative way, as groups of organisms that do not interbreed with one an-other in nature. It should be emphasized that the question is not just whether they can interbreed and produce fertile offspring, but whether they actually do so under natural conditions. But the distinc-tion is even more subtle than that. The fundamental question in rela-tion to reproductive isolation is not simply whether or not two groups interbreed, but whether sufficient gene flow occurs to cause the gene pools of the two populations to converge. If the two groups, despite occasional hybridization, retain their separate genetic identities and pursue independent evolutionary paths, they must be considered to be distinct species. Of course, if they are geographically isolated, the question of whether they are also reproductively isolated cannot be easily answered. A more positive biological species definition is that a species is a group of individuals sharing a common gene pool and bound together by bonds of mating and parentage.

In practical terms, the biological species definition leaves a great deal to be desired. Obviously, it cannot be used to make taxonomic decisions about fossils, nor is it useful in decisions about similar but geographically isolated populations. It can only be applied to contemporaneous, sympatric populations where it is possible to observe in the field whether or not they interbreed. Clearly, such a restriction greatly limits the usefulness of this type of definition. However, as Winston Churchill once said about democracy, the biological species definition looks bad only until you consider the alternatives.

If the individuals studied are fossils or dead preserved specimens, species of necessity are defined in terms of morphological traits. But because morphologically different individuals are often members of the same species (for example, in polymorphic or polytypic species), whereas morphologically similar individuals sometimes turn out to be reproductively isolated, the degree of morphological similarity can hardly be considered a foolproof criterion on which to base taxonomic decisions. Furthermore, if morphological criteria are applied, sooner or later a question will arise about how morphologically different two groups must be to justify calling them separate species, and the morphological species becomes little more than the subjective construct of a taxonomist. This point leads to another type of species definition, namely, that a species is simply what a competent taxonomist considers to be a good species. This approach has considerably more merit than might be evident at first glance, but leaving taxonomic decisions in the lap of the gods of taxonomy is the ultimate in the subjective definition of species.

The great advantage of the biological species definition, despite its limitations, is that it depends on the behavior and whims of the organisms involved rather than on the whims and prejudices of taxonomists. To apply the biological definition requires far more knowledge of the populations being studied than any other approach, but where it is used, the decisions are more apt to reflect biological reality than decisions reached by any other method. The great disadvantage of the biological species definition is that it is nondimensional, that it can only be applied to populations living together in the same place at the same time. It cannot be used with populations that are isolated from one another by either time or space.

Despite the limitations to all the various species definitions, in the past species were considered to have objective reality, and the goal of taxonomists was to describe each species as a natural biological unit. As long as species were regarded as fixed entities, this goal was considered attainable, and the difficulties in classifying species were attributable to the imperfection of human understanding of the complexities of nature rather than to the species themselves. Darwin threw a sizable monkey wrench into this perception of species, however, when he said that they were not constant but instead were the products of evolution.

Patterns of Evolution

Now we must consider how species evolve. The first and simplest way is called phyletic evolution or anagenesis. Phyletic evolution involves a single lineage changing through time, with the original species giving rise to a recognizably different descendant species at a later date. In this case, the species functions as the evolving unit, and there is a transformation through time in the characteristics of the evolving species. Although the end product is recognizably different from the original species, there is genetic continuity throughout, and it is impossible to pick any point and say that here species A stops and species B begins, unless the point is chosen quite arbitrarily or there is a convenient gap in the fossil record.

The mechanism for phyletic evolution is again quite clearly the same as that for evolutionary changes within a single deme and for the genetic divergence among demes that results in the formation of ecotypes and geographical races. In other words, phyletic evolution results from the combined effects of mutation, selection, and drift.

Another major pattern for the origin of species is known as cladogenesis or speciation in the restricted sense and involves the multiplication of species in space. In contrast to anagenesis, where a single species gives rise to just one descendant species, with cladogenesis a single species gives rise to two or more reproductively isolated, contemporary species. This pattern of evolution is not so simple as that of anagenesis, for it involves not just the origin of genetic differences between populations but also the origin of reproductive isolating mechanisms. The crucial step in the multiplication of species is the origin of isolating mechanisms between populations not previously reproductively isolated. Geographical or physical isolation serves to keep members of the same species from mating, but it is not an isolating mechanism. Isolating mechanisms are under genetic control and serve to prevent or reduce interbreeding between individuals even when they are sympatric. The problems with speciation then are twofold. What causes genetic divergence within a wide-ranging species, and how does a freely interbreeding ancestral species population become separated into two or more reproductively isolated groups?

The causes of genetic divergence within a species have already been discussed and seem as applicable to cladogenesis as to anagenesis. The origin of genetically controlled reproductive isolating mechanisms is another matter, however. Sexual reproduction involves a number of steps, and the process can be interrupted in a variety of ways. Isolating mechanisms have been classified (for discussion see Merrell, 1981, chapter 14) into premating and prezygotic isolation (ecological, temporal, and ethological isolation), postmating but prezygotic isolation (structural and physiological isolation), and postmating and postzygotic isolation (hybrid inviability, hybrid sterility, and hybrid breakdown). In each case, the isolating mechanism either prevents mating from occurring or, if mating occurs, blocks fertilization or, if fertilization occurs, affects the viability or fertility of the hybrid progeny. The net result is that normal, viable, fertile offspring are seldom produced from matings between members of reproductively isolated groups.

The genetic divergence that leads to the formation of ecotypes or subspecies does not necessarily produce reproductive isolation so that the origin of genetic isolating mechanisms may occur independently of the origin of polytypic variation. One theory for the origin of reproductive isolation is that it develops as an incidental by-product of genetic divergence (Muller, 1940, 1942). This theory is based on the implicit assumption of allopatric speciation. If geographically isolated populations diverge genetically from one another in terms of, say, habitat preference or breeding season or sexual behavior, then when they again become sympatric, these ecological or temporal or behavioral differences may be sufficient to form barriers to successful gene flow between the populations. Once this step has been taken, these two populations can coexist sympatrically and pursue their independent evolutionary paths, and two species, of common descent, will exist where only one existed before.

Another theory for the origin of reproductive isolation is that it is the product of natural selection. This theory also tacitly assumes allopatric speciation. If the occasional hybrid offspring formed from matings between individuals from two divergent populations are less fit than the offspring from intrapopulational matings, selection will act not only against the hybrids, but also against the behavior in the parents that led to hybridization, and thus ultimately against hybridization itself (Dobzhansky, 1940). This theory, of course, is complementary to the theory that reproductive isolation is an incidental by-product of genetic divergence. If any incipient isolation has started to develop during a period of spatial isolation, it can be reinforced by natural selection favoring homogamy over heterogamy when the populations again become sympatric.

In addition to anagenesis and cladogenesis, another pattern for the origin of species must be recognized, reticulate evolution, which shows a phylogenetic web rather than a phylogenetic tree. The clearest example of reticulate evolution is the origin of allopolyploid species discussed in chapter 6 as an example of saltational evolution. Here hybridization between two species followed by chromosome doubling gives rise at a single step to a new species that is reproductively isolated from both parental species. The reproductive isolation between the allopolyploid and its parents is chromosomal rather than genic, and no period of allopatric isolation is required. Reticulate evolution has clearly been of great importance in the evolution of the higher plants where up to half of all species are estimated to be allopolyploid in origin. Although allopolyploidy is the most clear-cut form of reticulate evolution, hybridization followed by selection among the hybrids even without polyploidy is a form of reticulate evolution found in both plants and animals, especially in disturbed habitats. Chromosomal reproductive isolation may result not only from polyploidy but also from chromosomal rearrangements such as inversions and translocations. The importance of reticulate evolution and of chromosomal reproductive isolation has undoubtedly been understated in the past because of evolutionary biologists' preoccupation with allopatric speciation as the primary generator of new species.

We have looked, very briefly, at polymorphic and polytypic patterns of variation, at the pattern of microevolutionary, infraspecific changes that give rise to ecotypes, geographical races, and clines, and finally at the nature of species and their patterns of origin: transformation through time, multiplication in space, and reticulate evolution. The patterns of evolution thus may be linear or branching or weblike; the lines of descent may join or separate or continue as a single lineage. These possibilities are not just alternatives, for within a given phylogeny, branching and anastomosis and simple linear evolution may all be going on simultaneously.

It is worth noting that Simpson used "mode" in his well-known classic, *Tempo and Mode in Evolution* (1944), for what we have been calling "patterns" of evolution. However, in the revision entitled *The Ma-*

jor Features of Evolution (1953), his usage changed, and he called these geometrical representations of evolution "patterns" rather than "modes" of evolution.

Modes of Evolution

Next we shall consider the modes of evolution, the ways in which species originate. At the outset, it should be realized that with so many millions of diverse species, past and present, it is highly unlikely that all of them will have originated in the same way. The mode of origin will be influenced by the environmental conditions in which the organisms live, by the way they are distributed in space and time, and by their genetic system. All of these factors will be acting simultaneously, of course, but initially we shall deal with them separately.

First, let us consider the role of geographical distribution in the origin of species. It is widely agreed that the most common mode of origin for species is geographic or allopatric speciation. Allopatric refers to individuals or populations that are spatially isolated from one another, usually with the added implication that they are so far apart that mating between them does not occur. The strongest advocate of allopatric speciation has been Mayr (1963), who has written (1970) that "geographic speciation is the almost exclusive mode of speciation among animals, and most likely the prevailing mode even in plants." Geographic speciation can be thought of as a continuation of the processes that give rise to ecotypes and subspecies. If geographically isolated populations of a species diverge genetically from one another to the point where they are not just geographically isolated but are reproductively isolated as well, they will have achieved the status of separate species, for even if they again become sympatric, the genetic isolating mechanisms will prevent gene flow between them and they will persist as independent evolutionary units.

Mayr's "founder principle" is simply another form of allopatric speciation in which the founder population is presumed to be much smaller than the parental species population. It has also been called "peripatric" speciation because the new species are thought to originate in small, isolated, peripheral populations. The genetic basis for the "founder effect," the so-called genetic revolution, is in reality nothing more than a combination of the effects of selection and genetic drift.

Although "parapatric" speciation has been postulated as a possible form of speciation in contiguous populations (Mayr, 1970), this concept is based on the assumption that the allopatric model of speciation

is the predominant mode of speciation and that the adjacent populations are in secondary contact after a period of isolation. It now appears, however, that many contiguous populations have never been
disjunct and have diverged while remaining in primary contact
(White, 1978; Grant, 1981). The difficulties with usage mentioned
earlier—that parapatric can be applied to contiguous populations but
not to individuals, which are either sympatric or allopatric—plus the
frequent difficulty in determining whether parapatric populations are
in primary or secondary contact, make the whole concept of parapatric speciation of dubious value.

We have already seen that sympatric speciation has been a major
mode of evolution in the higher plants where so many species are allopolyploid in origin. Another type of sympatric speciation, which involves chromosomal rearrangements, has been dubbed "stasipatric"
speciation by White et al. (1967; see also White, 1968). A similar mode
of evolution in plants was suggested earlier by Lewis (1953, 1962,
1966). The rationale for postulating stasipatric evolution is the high
incidence of karyotypic differences between closely related species of
plants and animals. White (1978) has estimated that only from 1 to at
most 10 percent of related species of eukaryote organisms have homosequential chromosomes. These estimates suggest that chromosomal rearrangements normally accompany the differentiation of species, and that the origin of species solely through genic divergence on
homosequential chromosomes is a rather rare exception.

Of course, one significant question is whether the chromosomal rearrangements play an integral part in the origin of reproductive isolation between the diverging populations or merely occur subsequently in populations already reproductively isolated by other
mechanisms. If the latter were the case, the rearrangements would be
simply an incidental by-product of the continuing phyletic evolution
of the populations. However, the evidence suggests that in many
cases the chromosomal rearrangements represent the initial and crucial step in the separation of species and that the genic differences between the populations evolve following this initial step (White, 1978).
In brief, the evidence consists of many genetically very similar species
that have widely different karyotypes, indicating that the chromosomal rearrangements leading to reproductive isolation and species
formation have occurred prior to any significant amount of genic divergence. Stasipatric speciation or evolution "in place" is used to describe this type of speciation because the chromosomal rearrangements are assumed to occur anywhere within the existing range of the
species, and, if they become established, they form an immediate iso-

lating mechanism between the parent and daughter populations. Thus stasipatric speciation is a case of both sympatric and saltational evolution.

Evolutionary Rates

From the geographical or spatial aspects of speciation, we must now turn our attention to evolutionary rates, to the tempo of evolution. As stressed earlier in chapter 8, there seems to be widespread agreement that evolution is gradual, that the rate of evolution is slow. However, we have just finished pointing out that both allopolyploidy and chromosomal rearrangements provide us with well-established mechanisms for the rapid, saltational origin of species. The old ideas of macroevolution or saltational evolution being mediated by macromutations or systemic mutations producing "hopeful monsters" have little evidence in their favor. However, a rapid, even instantaneous, origin of species can occur through either polyploidy or chromosomal rearrangements, both of which can give rise to reproductively isolated populations that are, by definition, new species even though in such cases the genic divergence often thought to characterize the formation of species follows rather than precedes the speciation event.

Nevertheless, the gradual evolution typical of the allopatric formation of species certainly accounts for the evolution of a great variety of species. In this case, the gradual accumulation of genetic differences precedes the achievement of reproductive isolation, although, as pointed out previously, dominant genes of major effect may play a more significant role in evolution than has been generally recognized and could speed up the rate of evolution beyond that possible with just genes of minor effect. In other words, limiting the concept of allopatric speciation to the very gradual evolution of quantitative traits controlled by numerous multiple factors may be too restrictive. Even the rate of allopatric evolution may get an occasional lift from the incorporation of a favorable dominant or partially dominant gene.

The concept of punctuated equilibrium formulated by Eldredge and Gould (1972), which deals with evolutionary rates, has drawn a surprising amount of attention, surprising because so little in the concept is new. In essence, they reported that the fossil record reveals long periods of stasis with little or no evolutionary change, punctuated by brief periods of rapid change during which speciation occurs. On the basis of this observation, they launched an attack on the modern synthetic theory of evolution, using Mayr's version of the modern synthesis (Gould, 1980, 1982b) as the basis for their attack. However, Mayr's concept of gradual allopatric speciation, as we have seen, does not adequately represent modern evolutionary theory.

The punctuationists have tended to stress the long periods of stasis in the fossil record as the most difficult to explain under current theories. It is worth pointing out that the reported stasis is morphological and is limited to only those morphological traits that happen to become fossilized. Paleontologists have very little information about the genetics, biochemistry, physiology, ecology, or behavior of the species they study, and thus have no way of knowing what evolutionary changes might have occurred that left no morphological trace. Thus, a finding of little or no change in shell dimensions, for example, provides only rather tenuous evidence for evolutionary stasis for all the other characteristics of the species. Nevertheless, the concept of punctuated equilibrium stimulated a great deal of discussion and empirical work.

Furthermore, the brief periods of rapid change found in the fossil record turned out to encompass tens or hundreds of thousands of years, more than adequate time for any one of a number of modes of evolution to come into play, including "gradual" evolution involving the accumulation of numerous small genetic changes. Therefore, the supposition that these "rapid" speciation events required some new and different mechanism of evolution, which led Gould to resurrect Goldschmidt's ideas on macroevolution (1982a), stemmed more from the different time scale used by paleontologists than from any flaws in the modern synthesis. Properly understood, the modern synthesis is a remarkably adaptable, useful, and resilient theory.

The truth of the matter is that just about any evolutionary rate, or any combination of evolutionary rates, can be readily interpreted under the modern synthesis. There is no rule that evolutionary change must inevitably occur. In a stable environment, the selection pressures will tend to remain the same and will minimize the effects of random factors for change such as mutation or drift and counteract the effects of gene flow. Thus stabilizing selection can readily explain evolutionary stasis, when and if it occurs.

Furthermore, there is no reason to suppose that evolutionary change will only occur at certain limited times, or that species will form in bursts of speciation. Given the proper conditions, such bursts may well occur, but they are a product of the prevailing biological and physical conditions, not of some sort of inherent drive to form species after a period of stasis. Lacking any sort of plausible evolutionary mechanism, punctuated equilibrium in its original form smacks too much of entelechy or mysticism to be taken seriously.

Rates of speciation and evolution may range from instantaneous, in the case of allopolyploid formation, to imperceptible, in the case of well-adapted species living in a stable physical and biological environment. The only caveat here is that adaptations are never perfect, so that selection will continue to refine the adaptation of a species to its environment, and environments are never completely stable through time so that selection pressures may gradually and imperceptibly change. For these reasons, it seems likely, even in species supposedly showing long-term stasis, that subtle evolutionary changes continue to occur even though they may not show up in the sketchy sample of morphological traits recorded in the fossils.

At this point, a few words about how to measure rates of evolution quantitatively may be in order. Simpson (1944) coined the terms bradytelic, horotelic, and tachytelic, for slow, medium, and fast evolutionary rates. The unit measured was usually the length of survival of extinct genera, with survival measured in millions of years. Different genera in the same group were compared for their duration in the fossil record. Those that survived for long periods of time were said to evolve slowly; those that endured for relatively short periods in the fossil record of the group compared to the general run of horotelic genera were said to have evolved rapidly. This comparative method probably provides some useful information about relative rates of change within a particular group of organisms. However, when comparisons are made between disparate groups (for example, pelecypods and mammals), the results become less reliable. One reason is that a genus, a group of closely related species, is the creation of a taxonomist, and given the different characteristics of pelecypods and mammals, the differences in the fossil records of pelecypods and mammals, and the differences between pelecypodian and mammalian taxonomists, it seems doubtful that a genus of pelecypods is equivalent to a genus of mammals. One can quibble as well about whether the survival time of a genus in the fossil record is even an appropriate measure of the rate of evolution. This type of evolutionary rate was called a taxonomic rate (Simpson, 1953).

Morphological evolutionary rates have also been determined. In such cases, the rate of change in a unit character or in a character complex, including groups of allometric traits, has been studied. A relative time scale based on stratigraphy or on the geological sequence is often used, though time may also be expressed in millions of years. But unless some reliable absolute method such as radiometric dating is available, the time scale always presents a problem, for estimates of absolute geological time based on correlative data are obviously subject to error.

Another nagging question about the time scale on which to measure evolutionary rates is whether absolute time is the appropriate scale at all. The basic processes of evolution, the changes in the kinds and frequencies of genes in populations, occur *between* generations, not within generations. Therefore, it seems more reasonable to measure evolutionary rates in terms of generations rather than in terms of elapsed time, although Nei (1987) asserts that neutral genes evolve in accordance with a time-based rate but selected loci do not. In theory, a species with a short generation length should be able to evolve more rapidly than one with a long generation time. The few attempts to test this theory have led to the conclusion that generation length is not a significant factor in the observed rates of evolution, and the question has been more or less swept under the rug. Given that the many other factors affecting evolutionary rates were not controlled in these studies, such a dismissal may be premature.

The Molecular Clock

More recently, a new approach to the study of evolutionary rates has been based on rates of molecular evolution. Earlier it was stated that evolutionary change is not inevitable, that stabilizing selection may maintain a population relatively unchanged under stable environmental conditions. Also, the concept of punctuated equilibrium involves periods of stasis, of little or no evolutionary change. However, the molecular clock hypothesis, resulting from studies of molecular evolution, seems to contradict the idea of evolutionary stasis or stability. This hypothesis states that the substitution of subunits in specific protein or nucleic acid sequences proceeds at a stochastically constant rate. These amino acid or base substitutions are assumed to be neutral in their effects on fitness and not to change the integrity of the evolving molecules. Because the replacement rate for neutral mutations equals the mutation rate per gene per generation, the image that emerges here is not one of stasis but of evolution chugging along at a steady rate based on the neutral mutation rate, but going nowhere as these mutations are adaptively neutral, or at least adaptively equivalent. Although Wilson, Carlson, and White (1977) wrote that "the discovery of the evolutionary clock stands out as the most significant result of research in molecular evolution," others have been somewhat less euphoric.

If the amino acid and nucleotide substitutions that occur during macromolecular evolution result from the chance fixation of selectively neutral mutations, the rate of fixation, which equals the neutral mutation rate per gene per generation, will be stochastically constant. That is, the rate of protein or nucleic acid evolution can serve as a molecular clock because the average rate of molecular evolution will be quite uniform over long periods of time. The rate is not a metronomic ticking like a grandfather clock, however, but is probabilistic like radioactive decay, where the probability of decay is constant but the actual events are stochastic. The validity of the molecular clock hypothesis and also of the neutrality theory of molecular evolution can be tested by plotting the average number of amino acid or nucleotide differences among a group of related species against their estimated time of divergence from one another. When such plots have been made, the number of substitutions appears to be directly proportional to the elapsed time since divergence, the points falling very close to the straight line expected for a uniform rate of substitution. Thus, all seems well, and the hypothesis seems to be supported by the data.

Nevertheless, some puzzling questions remain. For one thing, the theory predicts that the molecular evolutionary rate should be constant when time is measured in generations, but in these plots time was measured in years rather than generations. Since mutation rates in species as different in generation length as bacteria, *Drosophila*, mice, and humans are much more similar per generation than per year, generation length would seem to be a more appropriate time scale than years.

Another finding is that different molecules have different rates of evolution and that even different parts of the same molecule have different rates of evolution. The explanation offered is that some molecules and some parts of molecules are functionally more important than others, and that the faster rates of neutral evolution occur in the functionally less significant molecules or portions of molecules. But this sort of thinking implies selective constraints on the molecules, and suggests that only the less important molecules or portions of molecules can safely be used as molecular clocks.

Another assumption is that many of the substitutions do not lack adaptive significance but rather are adaptively equivalent. The so-called silent or synonymous mutations are those in which a base substitution in the DNA does not lead to an amino acid substitution in the protein because of the redundancy of the code. These silent mutations are assumed to be adaptively equivalent, a critical but perhaps unwarranted assumption, for different synonymous mutations may be differently constrained by the abundance of various nucleotides or tRNAs or in other ways. This assumption is logically shaky, for it deals with the subject being investigated, that is, with neutrality. More recently, the neutralist theory has included "nearly neutral" mutations because they will behave like neutral mutations in effectively small populations. One of the more troublesome aspects of the neutralist position, however, is the assumption that the various mutants have a constant fitness. There seems to be little appreciation of the possibility that a given molecular substitution may reduce fitness at one time, be neutral or nearly so at another time, or may even confer an adaptive advantage if conditions change. This, plus the impossibility of ever proving adaptive neutrality, given the nature of scientific proof, leaves room for considerable skepticism about the meaning of the molecular clock.

Even if one accepts the premises on which the molecular clock is based, it turns out that the variance in the rate of the molecular clock is more than twice as great as that expected if the clock were strictly stochastic. Another shortcoming of the theory is that in most cases the time of divergence from a common ancestor against which the clock is calibrated is obtained from the fossil record. Such geological ages are subject to considerable uncertainty, both about the interpretation of the fossil record itself and about the geological age of the common ancestor. Furthermore, the procedure smacks of circularity, for the clock is first calibrated against one measure of geological time and then is used as a different measure of geological time. It would also be comforting to know the actual amino acid or nucleotide sequences of the common ancestor that lived many millions of years ago, for the molecular clock hypothesis ignores gaps in sequences, considers only the minimum number of substitutions required to go from one sequence to another, and ignores the possibility of intraspecific polymorphic and polytypic variation.

The molecular clock hypothesis, based on neutral or nearly neutral mutations, leaves us with a clock that seems to run, if measured over long enough periods of time, at a rather constant but somewhat more variable rate than expected if mutation rates are constant. If molecules are evolving at fairly constant rates, then molecular evolution must proceed independently of adaptive evolution and evolutionary stasis must not hold, at least for molecular evolution. The idea that molecular evolution can be decoupled from morphological, physiological, behavioral, and other forms of adaptive evolution seems to fly in the face of logic, and the evidence from a mere recital of human molecular diseases (McKusick, 1988) should suffice to indicate the close relationship between molecules and fitness. If neutral molecular evolution only goes on in the adaptively less significant molecules or parts of molecules, which go their own merry evolutionary way, it must play only a minor role in evolution.

However, the application of Ockham's razor ("Essentia non sunt multiplicanda praeter necessitatem," i.e., the best explanation of an event is the simplest, the one requiring the fewest assumptions) suggests a more fruitful approach to molecular evolution. Just as it is impossible to prove the neutrality of certain molecular substitutions, so it would be impossible to prove the neutrality of certain minor morphological differences even though under many conditions these differences may have no significant selective effect. Morphological evolution is not assumed, however, to be governed by two different modes of evolution, adaptive and nonadaptive, and it seems equally unnecessary and unwise to split molecular evolution into selective and neutral modes, especially given the difficulties in trying to determine the variations in fitness between different molecular forms. The random as well as the adaptive aspects of the evolutionary process have been integrated into the evolutionary mechanism already outlined, and molecular evolution can fall under this rubric just as easily as the evolution of morphological, physiological, or other traits. Again it should be stressed that molecules are part of the phenotype of the individual and are governed by the same evolutionary processes that govern the evolution of the rest of the phenotype. To assume that molecular evolution marches to a different drummer from all other kinds of evolution, with the beat set by the ticking of a molecular clock, certainly violates Ockham's rule. The greatest weaknesses of the neutralist theory of molecular evolution are the assumption of fixed fitnesses for molecules and the impossibility of proving that certain molecules are, in fact, always neutral in their effects on fitness. (Further discussions of the molecular clock and neutralism and selectionism can be found in Merrell [1981], Kimura [1983], and Gillespie [1991].)

Our understanding of rates of evolution is crude at best. Studies of living species suggest that evolutionary changes may at times occur very rapidly. Studies of morphological changes in the fossil record indicate very slow, gradual changes spread out over millions of years. The interpretation of changes at the molecular level seems still in a state of flux, with opposing viewpoints struggling for acceptance or compromise. What is clear is that rates of evolution may be variable or steady and may range from very high to very low or even to zero, depending on the entities being considered. Nevertheless, it is safe to say that all of the rates measured, however crudely, can be interpreted within the framework of the modern synthesis.

The Genetic System

The next factor to consider in the discussion of modes of evolution is the genetic system, the mechanism for the organization and transmission of the hereditary material. The genetic system in prokaryotes is very simple, a strand or ring of DNA, but in eukaryotes it consists of a group of more complex structures, the chromosomes, with DNA associated with proteins. Like other phenotypic traits, the genetic system itself has been modified during the course of evolution. The evolution of chromosomes and of mitosis and meiosis seems to have provided a mechanism for the orderly replication, recombination, and distribution of the genetic material.

The most familiar genetic system is found in sexually reproducing diploid species with separate males and females and with homosequential chromosomes, but many other systems are known. For example, in some species the life cycle may include both asexual and sexual phases, whereas in others reproduction may be exclusively asexual. Some species reproduce sexually but do not have separate sexes, for sexual reproduction is carried on by hermaphrodites, individuals with both ovaries and testes. The mechanisms of sex determination and sexual differentiation are important aspects of the genetic system. The function of the genetic system can be regarded as the self-perpetuation of the genetic material. Because of the vicissitudes of existence and the changing nature of the environment, self-perpetuation seems to demand not just an unerring replication of existing genotypes but some provision for genetic variation as well. The varieties of asexual reproduction ensure, barring mutation, the exact replication of existing genotypes from one generation to the next. Sexual reproduction, on the other hand, has greater flexibility, for it permits, through recombination, the production of an array of genotypes showing greater or lesser degrees of resemblance to the parental types. In fact, the function of sexual reproduction seems to be primarily to provide a mechanism for generating and releasing an optimum amount of genetic variability. The variations in modes of reproduction become most meaningful when viewed from this perspective.

In species with homosequential chromosomes, recombination between genes on different chromosomes is made possible by Mendelian segregation and independent assortment. Crossing over, mediated by chiasmata between nonsister chromatids, permits recombination among genes on the same chromosome. As we have seen, however, relatively few species have the comparatively open recombination system associated with homosequential chromosomes. Chromosomal rearrangements (inversions, translocations, insertions, duplications, and deficiencies) are frequently observed as a part of the genetic variation of a species, in many cases at polymorphic frequencies. Considerable effort was devoted to the study of inversion polymorphism in *Drosophila pseudoobscura* and related species by Dobzhansky and his associates. Although other possibilities were explored, at present the primary role of the inversions appears to be to lock up favorable gene combinations within the inversions and to perpetuate heterosis in inversion heterozygotes. The effect of chromosome polymorphism in general appears to be to restrict the amount of recombination compared to the amount possible if all chromosomes were homosequential.

The differences in karyotype between closely related species are so common that chromosome rearrangement is thought to be a frequent concomitant of the origin of species. Even though other possibilities cannot be excluded, its most likely role is to contribute to the origin of reproductive isolation between these formerly freely interbreeding populations. Allopolyploidy in higher plants also acts as an isolating mechanism between the allopolyploid and its parents. The success of these new species may be related, however, to the fact that allopolyploidy provides a means of generating permanent heterosis as well as a developmental vigor and stability not possible in the corresponding diploids.

The way the genes are organized into chromosomes and distributed to subsequent generations of cells and individuals by mitosis and meiosis shows many variations from the basic mode of sexual reproduction by diploid individuals with separate sexes. The simplest interpretation of these variations in the genetic system is that the genetic system itself has been influenced by natural selection, which in this case seems to be regulating the amount of genetic recombination. As the amount of genetic recombination may have far more adaptive significance than the number of bristles on a fly's thorax, for example, there is no more reason to doubt that natural selection can regulate recombination than that it can affect the number of bristles on a fly.

Another aspect of the genetic system seemingly controlled by natural selection in natural populations is the mating system. The mating system determines the amount of inbreeding and outbreeding that occurs in a population, and this, in turn, influences the amount of genetic variation released each generation. The ultimate in inbreeding is asexual reproduction, for all the genotypes produced are exact replicas, barring mutation, of the parental genotype. Obligatory selfing in hermaphrodites is a form of inbreeding that rapidly produces homozygosity whereas any degree of crossing among hermaphrodites will slow the rate of increase in homozygosity. The most intense form of inbreeding in species with separate sexes is sib mating or parent-offspring mating, whereas mating among less closely related individuals will increase homozygosity at somewhat slower rates.

Sexual reproduction, in general, seems to be a means to ensure outbreeding. In many hermaphrodites, obligatory outcrossing is found because of self-sterility, different maturation times for the male and female gametes, or other devices to enforce crossing. The separation of the sexes, of course, is a form of obligatory outbreeding.

Between asexual reproduction and obligatory selfing on the one hand, and sexual reproduction between separate sexes on the other, many variations in mating systems have been found. The alternation of generations between asexual and sexual reproduction seems to combine the best of both worlds, asexual reproduction permitting the rapid increase in well-adapted genotypes under favorable conditions and sexual reproduction permitting the generation of a variety of adaptive experiments when conditions change. The most viable and reasonable explanation for the variety of mating systems seen in plants and animals is that the mating system itself is a product of evolution and has been shaped by evolutionary forces in the same way as morphological or biochemical traits.

Thus far, we have considered patterns of variation, both polymorphic and polytypic, patterns of intraspecific evolution in the origin of ecotypes, clines, and races, and patterns of speciation (linear, branching, and reticulate). The most plausible geographical modes of evolution are allopatric, sympatric, and stasipatric speciation, whereas gradual, microevolutionary modes of evolution as well as saltational, macroevolutionary modes are both indicated by the evidence from evolutionary rates. The genetic system, which controls the organization, replication, recombination, and transmission of the genetic material, is a product of evolution just like other phenotypic traits, and understanding the origin and function of the manifold variety of genetic systems will provide a penetrating look into the mechanism of evolution.

The Mechanism of Evolution

One of the more difficult aspects of studying the mechanism of evolution is that the evolving population is subject to so many influences affecting the course of its evolution. Trying to keep track of the effects of mutation, gene flow, and selection as well as determining the effective population size in order to estimate random drift requires more skill than a juggler. And this is only half the battle, for some assessment of the biotic and physical environment is also required for a full understanding of the conditions under which the population is evolving. At some point it becomes necessary to consider the whole picture, to consider the combined effects of all of these factors acting simultaneously on the population—in other words, to study evolution in action.

Two major viewpoints emerged from such studies. R. A. Fisher (1930) was the primary proponent of the idea that evolution was governed by mass selection, that evolution occurred gradually through the accumulation of numerous small genetic changes, that selection coefficients were very small (of the order of 0.01 or less), and that the entire species population comprised a large panmictic breeding population. The last assumption was based on the supposition that the fate of small, isolated populations was extinction or swamping by gene flow from larger populations, so that small populations could not play a significant role in the evolution of a species. The assumption of small, virtually unmeasurable selection coefficients stemmed from Fisher's belief that genes of major effect were always detrimental because they deviated so far from the adaptive norm that they could play no part in the evolutionary process. He also held that the evolution of a species was inevitable because it existed in a "deteriorating" environment and had to evolve continuously in order to remain well adapted. The use of "deteriorating" in this context is somewhat misleading, for what is implied is not that the physical environment is deteriorating but rather that the continuing evolution of other species makes it necessary for any species to continue to evolve merely to stay well adapted under the existing conditions. This phenomenon was later dubbed the "Red Queen" effect by Van Valen (1973) for the Red Queen in *Alice in Wonderland* who ran as fast as she could just to stay in the same place. As the earth is an open rather than a closed system, the environment is deteriorating only in a relative rather than an absolute sense.

In contrast to Fisher's mass selection concept was Sewall Wright's shifting balance theory of evolution (1977). Wright often claimed that his theory was misunderstood or misrepresented by others, but according to Provine (1986), Wright himself was responsible for at least some of the misunderstanding. In contrast to Fisher, who believed that the entire population of the species constituted the evolving unit, Wright believed that both population structure and population size played significant roles in the evolutionary process. He envisioned the species, not as one large panmictic breeding population, but as subdivided into a number of more or less isolated breeding populations or demes, the degree of isolation depending on the amount of gene flow among the populations. Each deme constituted a separate adaptive experiment and would pursue its own evolutionary path, influenced by local selection pressures and accumulating its own unique array of alleles owing to the effects of mutation and random drift. If any deme were exceptionally successful, its numbers would increase, and these more successful gene combinations would then spread to other populations by gene flow. In this sense, Wright's interdemic selection might be considered a form of group selection, but this interdemic selection is more an extension of individual selection than it is a case of group selection, which is usually considered to be for the benefit of the group rather than the individual and to be opposed to individual selection.

Not only population structure but population size were important to Wright's theory, for he postulated that in the subdivided species population some of the subpopulations would occasionally become so small that random genetic drift could become a significant factor in determining gene frequencies. It was on this point that some of the confusion arose. Fisher, with his strong selectionist stand, dismissed the possible importance of small populations and of random genetic drift. In his early writings, Wright sometimes gave the impression that drift alone was sufficient to explain the differences among the subpopulations, but he later claimed that he was misinterpreted and that he regarded drift, like mutation, as a factor in producing genetic divergence in the subpopulations on which selection then could act. There can be little doubt that Wright was later a strong selectionist, but there can also be little doubt that biologists heard his earlier message about drift clearly, for they frequently used drift as an explanation for otherwise inexplicable genetic divergence even when there was no evidence one way or the other.

The Adaptive Landscape

The role of drift was an integral part of another of Wright's innovations, the concept of an adaptive landscape. Under this concept, the adaptive landscape was like a topographical map with adaptive peaks and valleys, and sometimes saddles connecting one adaptive peak with another. Wright argued that Fisher's concept of mass selection was of limited evolutionary significance because once a large population had reached an adaptive peak, it would be trapped there, unable to escape to a higher adaptive peak. In contrast, among the numerous, small, semi-isolated populations that he postulated, each an adaptive experiment, one deme might occasionally, by random drift plus selection, escape across a saddle to a higher adaptive peak. Therefore, the population structure he envisioned was the most favorable for supporting ongoing evolutionary change.

The adaptive peak concept had so much heuristic appeal that the figure has been reproduced repeatedly by Wright and many others ever since it was first published in 1932. Biologists seemed to have an intuitive feel for it even though Provine (1986) has recently questioned whether any of them, including Wright, really understood the concept, using such harsh expressions as "unintelligible" and "meaningless" to describe it. Provine's criticism of the fitness surface or adaptive landscape is directed primarily at the fact that Wright actually used two concepts of the fitness surface interchangeably, one based on gene combinations and the other on gene frequencies. Provine argues that the surface based on gene combinations is "unintelligible," for it does not generate a surface at all, and that the two versions are "mathematically wholly incompatible" and are certainly not equivalent.

The graphical representation of evolution on an adaptive landscape was so clear and convincing that Wright apparently tried to tie it in with his mathematical treatments of evolution as a shift in genotype frequencies or gene frequencies. Each gene combination, however, supposedly represents a point on an individual fitness surface while the use of gene frequencies gives a representation of a populational fitness surface, an entirely different matter. Nonetheless, throughout his long career Wright appeared to switch back and forth between the two versions of the fitness surface (Provine, 1986), using the individual fitness surface to illustrate his qualitative shifting balance theory of evolution and the populational fitness surface in relation to his quantitative equations dealing with gene frequency change in populations. Given this situation, one wonders why the adaptive landscape remained unchallenged for so long. The most probable reasons are that biologists were so preoccupied with the contour lines indicating fitness that they did not pay much attention to the other axes, or else respect for Wright blinded them to the problems inherent in the concept.

Others have also used the simile in discussions of evolution. Templeton (1982), for example, suggested that Wright's adaptive landscape should be turned upside down so that the peaks became pits and the valleys became ridges, with gravity the moving force. The trouble with this analogy is that it suggests that adaptive evolution is inevitable, as easy as rolling downhill, and it is difficult to avoid thinking of an adaptive pit as an evolutionary sinkhole rather than as an evolutionary pinnacle of successful adaptation. Like those before him, Templeton shows no sign of being troubled by Wright's analogy.

Stimulated by Wright, the paleontologist Simpson (1944, 1953) adopted the idea of an adaptive landscape, with modifications. He thought in terms of adaptive zones or adaptive grids rather than of a topographical landscape, and more significantly, he used phenotypes rather than gene combinations or gene frequencies to characterize the population. Because selection is phenotypic, there is a more direct relation between fitness and the phenotype than between fitness and gene combinations or gene frequencies. It is significant that when Wright was challenged by Provine about his "unintelligible" version of the individual fitness surface, he responded that the only way to save his version of the fitness surface was to use phenotypic traits rather than gene combinations. Despite the criticisms, the adaptive landscape seems to be here to stay.

One mode of evolution that deserves further mention is the "founder principle" (Mayr, 1954, 1963, 1970) because it is referred to so often in discussions of speciation. We have already pointed out the similarity between the founder effect and Wright's concepts. Sometimes called peripatric speciation, the founder principle is merely a special case of allopatric speciation. The concept uses such expressions as "cohesion of the gene pool," "loss of genetic variability," "breakdown of genetic homeostasis," "relaxed selection pressure," "genetic instability," "genetic reorganization," "genetic environment," and the culmination, a "genetic revolution." Although these catchy but nebulous phrases are memorable, Templeton (1982) has pointed out that there are serious flaws in Mayr's conception of speciation in marginal populations through a founder effect. Foremost is that the founder effect and the genetic revolution are based on a faulty premise, for the evidence shows that founder events do not lead to a considerable loss of genetic variation nor to greatly increased levels of homozygosity.

Carson (1982 and earlier) has attempted a "reductionist" approach to the various modes of speciation, trying to find a genetic basis common to all. What has emerged is distinctively Carsonian, but appears to be in the direct line of descent from Fisher and Mayr: from Fisher because Carson believes that speciation of all kinds is mediated by reorganization of the polygenes, which he defines as any genes of small effect individually that contribute to continuous or quantitative phenotypic variation, and from Mayr because he believes that founder events result in disorganization of the polygenic system, which precedes reorganization and speciation. Although he explicitly recognized a debt to Wright's shifting balance theory, his reliance on numerous small polygenic changes is more reminiscent of Fisher than Wright. In his outlines of possible evolutionary factors, Wright (e.g., 1978, p. 461) always included the possibility that genes of major effect and chromosomal rearrangements could play a role in the evolutionary process, but otherwise he seemed to ignore them. Carson, on the other hand, specifically excluded White's ideas (1978) on the importance of chromosomal rearrangements in speciation so that on the whole Carson's adherence to evolution through numerous small genetic changes resembles Fisher's viewpoint more than Wright's. However, the founder effect is a far cry from mass selection.

The preceding discussion of patterns and modes of evolution should help to make clear that present-day evolutionary biologists differ not over whether or not evolution has occurred, which is how the creationists try to interpret their disputes, and usually not even over the mechanism of evolution, for which the theoretical framework was developed by the population geneticists, but rather over the relative importance of the various factors that bring about evolutionary change: mutation, selection, migration, and drift.

One noteworthy fact to emerge from the study of patterns and modes of evolution is how often one all-encompassing explanation has been sought for the origin and evolution of species. However, the members of living species range in size from single-celled microorganisms to giant sequoias and whales weighing tons. These wildly divergent species form a food chain or food web. Some, such as green plants, are known as producers; others, like the herbivores, function as consumers of the producers; and still others, the carnivores, are consumers of the consumers, while the decomposers tidy up after all the rest. The species in the food chain differ in size, in numbers, and in life history, with size usually increasing and numbers generally decreasing at each higher level of the food chain. Thus a pyramid of numbers is formed, with some species consisting of millions or even billions of very small individuals, whereas others such as top carnivores may consist of a few hundred large, widely scattered individuals. The genetic systems, methods of reproduction, and life histories of all these species are extremely diverse. Despite this diversity, Fisher and many others have held that evolution is gradual even though half of all higher plant species are allopolyploid and originated instantaneously. Mayr has adamantly insisted that geographic or allopatric speciation is the only significant mode of speciation. When Eldredge and Gould proposed their punctuated equilibrium mode of speciation, they seemed to think it made all other modes outmoded. To achieve a "reductionist" explanation for all modes of speciation, Carson had to exclude chromosomal rearrangements and allopolyploidy, both frequently associated with species formation. This reductionist, almost monotheistic approach to evolution and speciation is puzzling, for, given the number and diversity of living and extinct species, it seems highly improbable that one mode of evolution will fit all, and the evidence increasingly suggests that an explanatory pluralism is required to account for the multiplicity of ways that species have evolved.

Chapter 10
The Adaptive Seascape

The most vivid image to emerge from the study of evolution is that of the adaptive landscape, first proposed by Wright (1932) and later modified by Simpson (1944) and Templeton (1982). Although Provine (1986) criticized the concept on mathematical grounds, it is open to a more fundamental criticism. The features of a terrestrial landscape, with peaks and valleys, saddles and ridges, are fixed and unchanging, and thus the concept of an adaptive landscape carries the implication that the fitnesses of organisms are also constant and unchanging. However, fitness, the ability of organisms to survive and reproduce, is not a constant but is relative to the existing physical and biological conditions. Because environmental conditions are continually changing, relative fitnesses are in a constant state of flux. The adaptive surface, rather than being static and fixed, is dynamic and fluid. Organisms on an adaptive peak at one time may, as conditions change, find themselves in an adaptive trough. Although a fixed fitness for each genotype is easier to deal with conceptually and mathematically, such an oversimplification is remote from reality. As the physical and biological environment changes, the fitnesses of organisms vary accordingly. Therefore, the adaptive surface is best visualized as an adaptive seascape rather than an adaptive landscape, with the fitnesses of individuals rising and falling on the restless surface of the ocean as conditions change.

The Fitness Seascape

In the adaptive seascape, the vertical axis should be individual fitness, with the other axes based on phenotypic characters of the individual, which are more directly related to fitness than gene or genotype frequencies. In our earlier discussion of the phenotype, we saw that the phenotype is what we make of it, those particular traits out of an infinity of traits that we choose to score or to measure. Thus, like the landscapes based on genotypes or gene frequencies, the seascape, too, can be multidimensional, with as many dimensions as there are phenotypic traits measured and with each point on the adaptive surface representing a single phenotype and its fitness. But rather than having the fitness of each phenotype represented by a constant height, the fitness of the phenotypes will fluctuate as conditions change, and a phenotype at the peak of a fitness wave at one time may subsequently fall into a fitness trough. Thus as time passes, the fitness of a given phenotype may rise and fall on the adaptive surface depending on the immediately prevailing conditions. To the extent that environmental changes are cyclical (circadian, lunar, annual), the changes in the fitness surface are analogous to the waves of the sea.

The relation between the phenotype, the environment, and fitness has an added dimension of complexity, for not only does the environment change, but the phenotypes of organisms also change. In *Drosophila*, for example, the phenotype progresses from the fertilized egg through several larval stages to the pupa, which eventually metamorphoses into an adult fly. In the frog, the progression is from fertilized egg to tadpole and then through metamorphosis to the juvenile frog and finally to the sexually mature adult. Therefore, the phenotype, the entity on which selection acts, is not fixed but is constantly changing. Furthermore, many phenotypic changes are adaptive responses by the genotype to environmental stimuli, and the phenotype, the selection pressures, and the environment are all ephemeral, forming a complex interacting system.

The one constant in this situation is the genotype of the individual. Even though the genotype is expressed in a succession of phenotypes during development, the genotype itself remains fixed. We have stressed that fitness is a characteristic of the phenotype, but from an evolutionary standpoint what is important is the fate of the genotype.

Although theoretical population geneticists have generally assumed constant fitnesses to make their mathematics more tractable, such a simple assumption is quite unrealistic. However, the concept of variable fitness introduces a considerable complication into our thinking about the evolutionary process. Past practice has been to assign a fixed value to the fitness of a gene, a genotype, or a phenotype. This value is usually an average fitness or the product of all the separate fitness components, but it tells us very little about the actual selective trials and tribulations that an individual experiences during its

development. Some efforts have been made to treat selection in a fluctuating environment (Wright, 1940; Felsenstein, 1976; Gillespie, 1991,
chapter 4). For the most part, they have addressed the question of
whether such selection is a dispersive force leading to fixation of one
allele or whether it might support polymorphism.

Gene Expression during Development

Although the genotype of an organism is fixed, at any given time only
a fraction of the genome is being actively expressed (Browder, 1980).
A reasonable supposition is that different systems of genes regulate
the differentiation of the various developmental stages and that these
are called into play sequentially during the course of development. As
each portion of the genome comes to expression, it is exposed to selection, and the genotype and the series of phenotypes it engenders
during development can be thought of as threading their way
through an environmental minefield. If any stage of development is
not adapted to the existing conditions, it will be eliminated by natural
selection, and that organism, and its corresponding genotype, will be
lost from the population. Thus selection is continuously monitoring
the developing organism, which must be adapted to its environment
throughout its life. Moreover, not only may the fitness of a developing organism vary during the course of its development, but under
different sequences of environmental conditions, the same genotype
may generate different sequences of phenotypes and different patterns of fitness.

Obviously, the much maligned adaptive paradigm still has considerable viability. There is no stage in the life cycle when an organism
can be poorly adapted, no period when natural selection takes a va-
cation. Even the so-called adaptively neutral genes are not truly adaptively neutral, for they are adaptively equivalent to alleles of similar
effect, which is quite different from complete neutrality. The important thing about these alleles is that they are functional; their evolutionary significance stems not from their adaptive equivalence to one
another but from their functional role in the organism.

In the controversy over neutralism and selectionism, efforts have
sometimes been made to test for adaptive neutrality. Because of the
nature of scientific proof, such tests are, in fact, tests of adaptive significance. Failure to find adaptive significance may be regarded as evidence for neutrality, but given that only a small fraction of loci are
active at any particular time during development and that some loci
may only assume adaptive significance every few generations in

times of stress (e.g., from disease or famine), a failure to demonstrate adaptive significance is, at best, rather tenuous evidence for adaptive neutrality. Even the so-called neutral alleles may seem equivalent only because their differences in fitness are so difficult to detect.

We have said that fitness is a property of phenotypes but that phenotypes, like environments, are ephemeral. The genotype of an individual, barring somatic mutation, is constant. In sexually reproducing populations, however, the genotype of each individual, except identical twins, is unique, for genetic recombination reshuffles and deals out new combinations of genes each generation. Thus, genotypes too are ephemeral, lacking continuity. The entities that have continuity from generation to generation are the genes themselves and the gene pools of the breeding populations. These are the entities that undergo evolution, and that is why we attempt to measure evolutionary change by following changes in the genetic composition and gene frequencies of breeding populations. Selection and migration do not, however, act directly on genes or genotypes but operate indirectly through phenotypes to bring about evolutionary change. It is phenotypes that migrate and phenotypes that are exposed to selection. Although it is convenient to speak of selection favoring this or that gene, it bears repeating that, strictly speaking, selection does not favor genes or genotypes, but rather favors phenotypes generated by those genes or genotypes interacting with a particular set of environmental conditions. On the other hand, mutation does alter genes directly, and the random drift of gene frequencies associated with sexual reproduction in small populations is not directly dependent on the phenotypes of the organisms.

Variability and Abundance

The idea that an organism must be adapted at all stages of its life cycle runs contrary to ideas promoted by Fisher and Ford, Carson, and Mayr. Fisher and Ford (1928) reported that abundant species have more genetic variability than rare species and interpreted this to mean that variability increases as numbers increase because selection is relaxed under favorable environmental conditions (Ford, 1975). Carson (1975) also believes that selection is relaxed during what he termed population flushes. Oddly, he assumes that selection is also absent when a population crashes, with survival a matter of chance rather than of genotypic value, but it seems more likely that selection during a crash is apt to be particularly stringent. In fact, Carson believes that there is no selection during the entire flush-crash-founder cycle. An added corollary is that when selection is relaxed, disorganization of the genome into a "sundered and shattered gene pool" (Carson, 1982) occurs, followed by the reorganization of a balanced, polygenic genetic system molded by natural selection over hundreds or thousands of generations. Carson (1971) even went so far as to state explicitly that during this process "there must be an original condition in which the population is characterized by individuals which are not adapted to a relevant environmental state." Mayr (1954, 1963, 1970) also assumed that selection pressure is relaxed in isolated founder populations, but in this case the selection is relaxed in small populations rather than large.

The Founder Effect

In the previous chapter we mentioned some of Mayr's ideas on the origin of species associated with the founder effect and Carson's related ideas associated with the flush-crash-founder population cycle. Very briefly, Mayr believes that the gene complex of a species is delicately tuned by natural selection into a complicated system of interacting genes, that each species has a "cohesive" gene pool molded by natural selection into a "coadapted" genetic system. Coadaptation refers not to adaptation to external conditions but to internal genetic cohesion based on allelic and epistatic genic interactions. The small size of founder populations supposedly leads to a loss of genetic variability, an increase in homozygosity, and a breakdown in genetic cohesion or "genetic homeostasis." This genetic instability plus "relaxed selection pressure" makes possible a major transformation in the gene pool or "genetic revolution" and leads to the formation of new species.

Homoselection and Heteroselection

Carson has similar but somewhat more advanced ideas about the origin of species and acknowledges his debt to Mayr (1982) in the genesis of his views. He too believes that a species has a genetically cohesive, balanced gene pool, using Mather's (1941) concept of "internal balance" to refer to intrachromosomal epistatic interactions, and "relational balance" to refer to interchromosomal allelic and epistatic interactions. Because some species have only homosequential chromosomes and others are normally heterosequential (that is, are polymorphic for a number of chromosome arrangements), Carson postulated that there are open and closed genetic systems determined by the nature of the karyotype. In the homosequential parts of the karyotype, the genetic system is open because genetic recombination can occur freely. When chromosomal rearrangements such as inversions and translocations are present at polymorphic frequencies, duplications and deficiencies are generated in chromosomal heterozygotes during meiosis. The genes within a rearrangement form a closed genetic system because they are no longer able to recombine freely with genes on homologous chromosomes and tend to be locked up in a single package called a "supergene."

Chromosomal polymorphism tends to be high at the center of the range of *Drosophila* species and rare or absent in peripheral populations. Carson felt that the open genetic system of the homosequential peripheral populations, which are chromosomally homozygous but not genetically homozygous, permits these peripheral populations to respond adaptively through genetic recombination to local environmental conditions, forming the basis for geographical and clinal genetic variability. This form of selection, which gives rise to ecotypes and subspecies, was called "homoselection."

On the other hand, "heteroselection" prevails in those parts of the genome locked into a closed genetic system. In such a system, selection for strong internal epistatic interactions causes a block of linked genes in a chromosome segment to build up obligatory *cis* relationships, forming a coadapted chromosome segment or supergene. Moreover, in addition to building up internal balance within a chromosome segment, heteroselection may also act to build up relational balance between chromosomes. The relational balance is expressed as heterosis, with the heterozygotes for these chromosome segments more vigorous or otherwise superior to the corresponding homozygotes.

Coadaptation

Although the concept of coadaptation has won wide acceptance, the evidence for it is surprisingly weak. The whole concept is based on the idea that intrademic selection will build up coadapted gene complexes that interact favorably to produce maximum fitness. Efforts to demonstrate coadaptation have repeatedly revealed, however, that interlocality hybrids with sets of chromosomes derived from two different breeding populations are much more fit than intralocality hybrids in which both chromosome sets come from the same breeding population. (For further discussion of these experiments by Wallace, Dobzhansky, and others, see Merrell, 1981.) This result is directly contrary to what is expected if selection for "genetic cohesion," "genetic homeostasis," "genetic coadaptation," and a "balanced gene pool" actually occurs. If the theory is correct, the fine tuning of the genetic system to generate internal and relational balance within and between chromosomes depends on selection within a single breeding population. The awkward finding of maximum fitness in interpopulational hybrids casts serious doubt on the entire concept.

Moreover, plant and animal breeders have long sought to select for specific combining ability, that is, for chromosome sets that, when combined in hybrids, specifically complement one another to provide maximum yield. Selection for specific combining ability is thus equivalent to selection for coadaptation. A major effort in such selection is to select for overdominance as originally defined, that is, to select for allelic interaction at a single locus such that the heterozygote is superior to the two corresponding homozygotes at that locus. The experimental difficulties in such a test are formidable because the rest of the genotype against which this locus is being tested must be held constant. The evidence indicates, however, that selection for specific combining ability and overdominance has not been successful (Moll, Lindsey, and Robinson, 1964; see also Comstock, 1977; Falconer, 1977; and Gardner, 1977). Although heterosis is observed in the experiments, it turns out not to be true overdominance resulting from single locus heterosis but "associative overdominance" (Frydenberg, 1963) with the heterosis stemming from the retention of heterozygosity over short chromosome segments with the favorable genes in repulsion. This phenomenon has been blessed with several names in addition to associative overdominance: "pseudo-overdominance" (Mangelsdorf, 1952), "multilocus heterosis" (Hexter, 1955), and "apparent overdominance by repulsion linkage" (Wright, 1977). Thus, the idea that favorable complementary allelic interactions could serve as the basis for coadaptation seems not to be supported by the evidence.

The discovery of enormous amounts of protein polymorphism in natural populations is another factor to pose problems for the concept of genetic cohesion and coadaptation. If genotypes are finely tuned to form complex interacting systems, why is the gene pool cluttered with such an array of variants? The greater the number of variants, the less likely it becomes that they are all part of an integrated genetic system.

As pointed out earlier, only a fraction of the genome is active at a given time. The question then becomes, how can selection produce an integrated genome if only part of the genome is exposed to selection at any one time? When a symphony orchestra tunes up, all the instruments play at once, but if only a few tune up at a time, getting them all in tune becomes virtually impossible. By the same token, the discovery that most of the genome is inactive most of the time makes the concept of selection for genetic cohesion rather difficult to accept.

Coadaptation is postulated to involve the internal and relational balance of the genes within the genome, but the relation between the genes and the external environment is ignored. Because genetic cohesion and coadaptation have been closely tied to the concept of heterosis, and heterozygous advantage is assumed to hold in all environments, relatively little thought has been given to the relation between the genome and the environment in this context.

Given that the control of differentiation and development in individual organisms suggests the presence of integrated genetic systems, the nature of these systems needs to be explored. The concepts of coadaptation and specific combining ability are inadequate, however, to account for the origin of this sort of integration. The facts suggest instead that selection acts to produce what has been termed "general combining ability." The genes are tested, not in one particular gene complex, but in an ever-changing array of gene complexes and in a variety of environmental conditions. Those that survive this constant winnowing process persist in the gene pool of the population; the others fall along the way. The chromosomal polymorphism that is so often observed in natural populations seems to be preserved by natural selection, not to create an integrated, cohesive genetic system, but to establish permanent linkage disequilibrium and thus maximize and perpetuate the heterosis based on associative overdominance.

Hard and Soft Selection

The idea that selection is relaxed in small isolated populations or during flush-crash-founder cycles fails to make a distinction between what Wallace (1981) has termed "hard" and "soft" selection. Hard selection is both frequency- and density-independent while soft selection is frequency- and density-dependent. As pointed out previously, there are many genetic conditions that are detrimental under any (except remedial) environmental conditions. Hard selection will tend to eliminate them at all times, no matter whether the population is large or small or the conditions favorable or unfavorable. Soft selection, on the other hand, only comes into play as the population size approaches the carrying capacity of the environment. Even though soft selection may not be a factor when numbers are small or conditions favorable, hard selection will operate under all conditions and will maintain the population in an adapted state at all times. To suggest that selection is sometimes relaxed or that a population is sometimes maladapted represents a failure to recognize that hard selection will always maintain a basic level of adaptation.

For example, the absence of an esophagus in feeding organisms will be detrimental under any set of environmental conditions. This is an example of hard selection. Soft selection only acts, according to Wallace, when competition for available resources becomes stringent as the population approaches the carrying capacity of the environment. If there is sufficient food to support the development of 100 individuals, but only 50 are present, all can develop without difficulty, but if there are 1,000 individuals, soft selection will permit only 100 (or none) to develop successfully.

Between hard selection and soft selection, however, there is natural selection in the usual sense, which seems to have been overlooked in this scheme. For instance, even though there may be adequate food for the normal development of 100 individuals and only 50 are present, if these individuals vary in their ability to utilize the food, the rates of development of the variants may differ and one type may have a selective advantage over the others. Moreover, if the available supplies of food are different for different populations, one variant may be favored in one population and others in other populations, leading to divergence.

Adaptive Peaks and Valleys

Just as an individual organism must of necessity be adapted to its environment or it will be quickly eliminated by natural selection, a natural population must be adapted to its environment, that is, be able to survive and reproduce there, or it will soon cease to exist. This necessity puts a different light on such familiar evolutionary metaphors as Simpson's adaptive grid (1944, 1953) and Wright's adaptive peaks and valleys (1932, 1977). Simpson's adaptive grid consists of adaptive zones separated by "adaptive discontinuities" or "unstable ecological zones" while Wright's adaptive peaks represent areas on the adaptive topography where the organisms are well adapted, and the valleys areas of poor adaptation. The implication in both cases is that, for evolution to occur, a population must somehow manage to cross an adaptive discontinuity (or nonadaptive zone) from one adaptive zone to another (Simpson) or else cross a valley, an area of low adaptation, to move from one adaptive peak to another (Wright).

The difficulty with these concepts is that they suggest that in making a transition from, say, browsing to grazing, or from a plantigrade to a digitigrade mode of locomotion, the populations involved must have endured a period during which they were poorly adapted. It is improbable, however, that populations evolving from browsing to grazing starved during the transition because they lacked suitable food; if so, they would simply have become extinct. Similarly, it seems unlikely that there was a period during the transition from plantigrade to digitigrade locomotion when the animals involved had to hobble around until the evolutionary change was completed.

The assumption of relaxed selection pressure by Carson and Mayr as a necessary condition for their theories of the origin of species is an attempt to circumvent the awkward problem posed by Simpson's nonadaptive zones and Wright's valleys of low fitness. Closely related species can easily be visualized as occupying adjacent adaptive peaks on an adaptive landscape. Presumably at some time in the past their ancestors occupied the same adaptive peak. A major problem in evolutionary biology is to determine how they managed to move to separate peaks. Since normalizing or stabilizing selection prevents movement by a population down an adaptive slope, an easy way out is to assume that natural selection ceases to operate during the crucial period when the shift from one peak to another occurs. Thus, to resolve what is otherwise an awkward dilemma, dedicated selectionists like Mayr and Carson assume relaxed selection pressure at the most crucial point in their theories. Whereas the neutralists simply thought that, compared to stochastic events, selection was never very important to evolution, in this case the selectionists postulated a convenient but temporary cessation of selection.

The concept of an adaptive seascape offers an alternative to the dilemma posed by the nonadaptive zones on Simpson's adaptive grid or the valleys of low adaptation on Wright's adaptive landscape. If the adaptive peaks are not fixed but can move across the surface of the sea, a population could ride the crest of an adaptive wave from one adaptive mode to another without ever being forced to go through a poorly adapted period. Similarly, if the adaptive zone is not regarded as fixed, it becomes easier to visualize the transition from browsing to grazing by a population continuously well adapted even during the time when it forages by some combination of browsing and grazing. It is worth noting that even though Simpson (1953, p. 157) and Wright (1949, p. 380) both explicitly recognized the evolutionary impact of changing environmental conditions, they continued to deal with adaptive zones and adaptive peaks as if they were static features on the adaptive surface.

Adaptation in Leopard Frogs

Anyone who has worked with natural populations in the field, especially in Minnesota, is painfully aware of how changeable the environment can be and how much closer an adaptive seascape is to reality than an adaptive landscape. To add a touch of realism to the imagery, let us consider the life and hard times of the leopard frog, *Rana pipiens*, in Minnesota.

The state of Minnesota, larger than all of New England plus New Jersey, Delaware, and Maryland, rests on the forty-fifth parallel halfway between the equator and the North Pole. Although the state stays put, its climate changes annually from tropical to arctic and back again. Moreover, it is possible to track the steady progression of the seasons by following the daily averages in the weather data, but the day-to-day weather is so changeable that averages become almost meaningless. Although it is probably safe to say that snow is unlikely in July, long-range prediction is otherwise a dubious venture. To give just one dimension of the environment, the temperatures range from over 100°F in summer to less than−25°F in winter, with circadian variations sometimes as great as 40°F. Occasionally in winter, the temperature may never rise above freezing from mid-December to mid-February. Moreover, precipitation may range from cloudbursts dropping 10 to 12 inches of rain in a few hours to droughts lasting for weeks or even months. Faced with such a climate, what is the leopard frog, a poor amphibious ectotherm, to do?

Unlike many species of birds, leopard frogs cannot migrate south for the winter, but they do migrate to the larger rivers and lakes that are unlikely to freeze to the bottom. There the adult frogs spend nearly half a year, not dormant but quiescent, not "burrowed in the mud" where they would die from lack of oxygen, but snuggled in vegetation or debris on the lake bottom or along the river banks (Merrell, 1977). In early spring the sexually mature adults migrate overland to temporary ponds for the brief, intense breeding season and then disperse widely to meadows and other terrestrial habitats to gorge themselves on insects and other prey during the summer. In the fall they move back to the lakes and rivers to repeat the cycle.

This brief recital of the life history of the adult leopard frog brings out the variety of environmental conditions the frogs face each year.

For nearly half a year they must be adapted to survive in a cold, dark, aquatic habitat. Then in the spring, the adults leave this aquatic habitat during a period of extreme temperature fluctuations to seek out ponds suitable for breeding. This migratory behavior is succeeded by breeding behavior adapted to securing a mate and leaving offspring. Feeding and surviving in the warm summer terrestrial habitat place a whole new set of demands on the adaptive abilities of the frogs. Finally, in autumn the frogs must be able to find a body of water suitable for overwintering. When we realize that the genotype of the terrestrial, lung-breathing adult frog has already been phenotypically expressed as a developing embryo in the egg mass, as an aquatic, gill-breathing tadpole, as a metamorphosing tadpole, and as a juvenile frog, and that each of these stages is subjected to its own set of environmental conditions and selection pressures, it is clear that the leopard frog must have an adaptively versatile genotype to meet the vicissitudes of life and love in a cold climate.

Given the variation in the environment and in the phenotype of the individual, how can one measure fitness in frogs? Each individual has its own unique genotype and phenotype and fitness. In most natural populations, it is very difficult to determine the fitness of individuals because this requires a measure of their reproductive success, which in turn requires the identification of individuals and their progeny. Since in most species, it is impossible to identify and follow individuals throughout their lives, let alone identify their progeny among the members of the next generation, this poses a major problem. Therefore, indirect methods of assessment of fitness have been sought, for example, trying to estimate the average effect on fitness of a phenotypic trait or a gene in a group of individuals sharing the trait or the gene.

When one tries to assess fitness in this way, another problem rears its ugly head. Although the numbers of leopard frogs may fluctuate greatly from year to year, nevertheless within rather wide limits the population size appears to remain constant. There is no indication that the population of frogs is increasing or, despite some rumblings to the contrary (Gibbs, Nace, and Emmons, 1971; Maugh, 1972; Anonymous, 1973), decreasing. Even though a single *Rana pipiens* female in Minnesota produces from 2,000 to 5,000 fertilized eggs each breeding season, the state is not blanketed with frogs. If population size remains constant, however, only two of the thousands of zygotes produced by each female during her life can, on average, survive to sexual maturity. Obviously, the death rate must be extremely high.

The fact is that leopard frogs meet their end in an amazing variety of ways. The tadpoles may be sucked dry by dytiscid water beetles and the adults by leeches. The adults are prey to garter snakes in the summer, to fish in winter and summer, and to a host of other predators such as herons, skunks, and raccoons, to name just a few. They are also subject to bacterial, viral, and fungal diseases. Moreover, given the chance, they will eat each other, for adults have been captured with the feet of young frogs protruding from their mouths. In this cacophony of death, many of these deaths must be random. A major question, for one interested in fitness, is what proportion of the deaths is random and what proportion selective? The answer to this question is not quite so simple as one might wish.

If mating patterns in humans are studied, it is possible to show that mating is selective, or at least assortative, with respect to height, but random with respect to the MN blood group. Thus mating may be both random and selective at the same time, depending on which traits are examined. In similar fashion, if one tries to determine whether deaths in frogs are random or selective, the answer may depend on the particular traits chosen for study and the environment in which the study is conducted. Many of the deaths observed may appear to be random only because the observer has chosen to study factors unrelated to the current selective agents. For instance, if one is studying the effects of temperature on fitness at a time when predation by garter snakes is the major selective agent, one may conclude that the deaths are random with respect to temperature even though at other times in the life cycle temperature may be a significant selective factor. Similarly, the MN blood group in humans may not always be neutral with respect to fitness if the alleles affect components of fitness other than mating behavior.

Furthermore, if a large percentage of deaths is random at a particular time, it may be very difficult to detect the selective deaths among the welter of random deaths. If 99 percent of the deaths during a given period are random, for example, but 1 percent reflect selective differences, it might be extremely difficult to detect this difference even though, in the long run, this selection pressure would have an impact on the population.

Another complicating factor is that a selection pressure may exist for only a brief period but may have long-lasting effects on the genetic makeup of a population. It is sometimes stated, for example, that food cannot be a limiting factor in a population because the food supply always seems more than sufficient for the existing population. However, if a food shortage for gray squirrels during the winter is of very brief duration, it may have drastic effects on the population but be quite difficult to detect. The reason is that the squirrels will not all simultaneously keel over from starvation, but will become more vulnerable to predators and parasites, to loss of suitable nesting sites, and the like, which will lead to their demise. Since predators and scavengers are very efficient in disposing of the remains of dead animals, it is usually difficult even to find their carcasses. The animals simply disappear from the area under observation, and it is unknown whether they have died or simply dispersed to a more congenial locality. If during the brief period of food shortage, some of the squirrels are better adapted to survive than others, dramatic changes in the genetic composition of the population may occur, but a summer ecologist would remain totally unaware of the true cause of the change. Traits that have a very large selective value during this brief period might appear to be adaptively neutral at all other times. A similar argument can be made with respect to epidemics. A disease may sweep through a population only sporadically, wiping out large numbers of individuals, but its genetic effects will remain unknown unless the population is being monitored before, during, and after the epidemic.

Efforts to determine fitness in natural populations of frogs at times resemble the pursuit of a will-o'-the wisp. Traits can be roughly classified as biochemical, physiological, morphological, behavioral, and ecological. One may then try to determine the effects on fitness of differences in blood type, or temperature tolerance, or relative length of appendages, or attraction to blue light, or choice of overwintering site. All of these diverse traits appear to have an effect on fitness at some time in the lives of the frogs, for all are related to the adaptability of the frogs to their environment. Nevertheless, it may be very difficult to provide a convincing demonstration of their adaptive significance.

The well-adapted organism is adapted to its physical and biological environment at all levels from biochemical to ecological. These adaptations are hardly matters of chance. They have arisen through the action of natural selection and are constantly monitored and refined through the continuing action of natural selection. The selective process is so complex, however, acting on the phenotype simultaneously at so many different levels and on so many different traits, that the study of the effect on fitness of a single trait difference is highly artificial. Given these complexities, the difficulties in demonstrating the adaptive significance of particular traits should not be surprising. Failure to find adaptive significance is no proof that a trait lacks adaptive value. Although it is tempting to substitute the neutral theory of evolution for the adaptive paradigm because the neutral theory eliminates the need to study adaptive values, the neutral theory is the easy way out. All of the stages in the life cycle of the leopard frog are adapted to survival in a changeable and often hostile environment. This complex developmental history is hardly the product of a series of chance events. The neutral theory of evolution may explain some of the noise in the adaptive systems we observe, but it utterly fails to explain the evolution of adaptation itself, no matter whether in bacteria, fruit flies, frogs, or humans. The difficulties with the adaptive paradigm arise because of the simplistic way in which it is so often presented, not because of flaws in the concept itself.

Polymorphism in Leopard Frogs

To illustrate some of the problems related to the study of adaptation, let us consider two genetic polymorphisms in the leopard frog populations of Minnesota. Some frogs in Minnesota were so different from the typical spotted leopard frog, *Rana pipiens*, that they were at first tentatively described by Weed (1922) as two separate species, *Rana burnsi* and *Rana kandiyohi*. In Burnsi frogs, the spots are rare or absent, but in Kandiyohi, added pigment between the spots gives the frogs a mottled appearance. Crosses of Burnsi with typical spotted wild-type pipiens showed that the unspotted Burnsi condition is due to a simple dominant gene (Moore, 1942). Similar crosses between Kandiyohi and pipiens showed that the mottled Kandiyohi pattern is controlled by another dominant gene (Volpe, 1955). Despite their distinctive appearance, Burnsi and Kandiyohi are not species or even subspecies; they simply result from the effects of two different dominant genes in *Rana pipiens*.

An extremely rare type of frog was also found, which was phenotypically intermediate between Burnsi and Kandiyohi but strikingly different from both in being mottled but unspotted. When crossed with wild-type pipiens, these mottled frogs produced four kinds of progeny in equal numbers: pipiens, Burnsi, Kandiyohi, and unspotted mottled frogs like the rare parent. This crucial cross demonstrated that these rare frogs were double dominant heterozygotes for both Burnsi and Kandiyohi and that, even though both dominants affect the pigment pattern, they are neither allelic nor linked but occur at separate loci on different chromosome pairs (Volpe, 1960; Merrell, 1972).

Since Burnsi and Kandiyohi were not separate species but were produced by simple dominants, the intriguing question became, why are these genes so common in Minnesota populations of *Rana pipiens*? In other words, what, if any, is the adaptive significance of these dominant mutations in natural populations of leopard frogs? Given the possible complexities in the study of adaptation, it is helpful if the genetics are simple. Because Burnsi and Kandiyohi are each controlled by a single dominant, the genetics of these traits is the simplest possible, for if an individual carries one of these genes, it is expressed in its phenotype. Ordinarily, wild-type alleles are dominant, but in these cases the wild-type alleles that produce the spotted pipiens phenotype are recessive to their rare dominant Burnsi and Kandiyohi alleles.

At the outset of the work the only information available was based on fewer than 300 leopard frogs collected in Minnesota (Breckenridge, 1941). So little was known about Burnsi and Kandiyohi that the first task was to learn something about their frequency and geographical distribution.

Field data were eventually gathered on more than 30,000 frogs. The distribution of Burnsi appeared to cover about 100,000 square miles in Minnesota, western Wisconsin, a bit of northern Iowa, and the prairie lake country of the eastern Dakotas. Although 30,000 may seem like, and was, a lot of frogs, it still represents a sample of only 1 frog per 3 square miles over a period of more than a decade, and thus constitutes but a tiny fraction of the total frog population in the area during that time. Kandiyohi was somewhat more limited in its distribution, being confined to the prairie region of western Minnesota and the eastern Dakotas. The range of the *Rana pipiens* species complex covers most of North America, so that, even though these genes are common over some 100,000 square miles, their distribution involves only a small fraction of the total range of the species complex.

The phenotypic frequency of Burnsi frogs in the samples from wild populations was occasionally as high as 10 percent, but typically ranged from 1 or 2 to 5 percent. As this was a relatively rare dominant, the gene frequency was therefore about half as great, that is, from 0.5 to 5 percent. The frequency of Kandiyohi frogs was similar to Burnsi in that it ranged from about 1 or 2 percent to 5 percent, with the gene frequency again about half as great. Although rather low, such frequencies are too great to be maintained by recurrent mutation and must represent some form of balanced polymorphism. One of the more puzzling aspects of the data was that mutants with such low frequencies were so widely and uniformly distributed over such large areas. The highest frequencies of Burnsi were found in an area known as the Anoka sand plain in east-central Minnesota, and as noted,

Kandiyohi was confined to the prairie region of western Minnesota and the eastern Dakotas.

At this point, a few words about the geological history of the area are in order. Most of the terrain where the mutants are found is dominated by the effects of recent glaciation during the Wisconsin age, which ended just 10,000 years ago. The last advance of the Wisconsin ice covered much of Minnesota with ice thousands of feet thick extending as far south as Des Moines, Iowa. Within the area formerly covered by the Wisconsin ice are the thousands of lakes, ponds, and swamps characteristic of the region, but beyond the limits of the most recent Wisconsin ice, the land is well drained, and lakes and ponds are rare.

From the Des Moines lobe of the Wisconsin ice, a sublobe branching off to the northeast as far as Grantsburg, Wisconsin, diverted the Mississippi River far to the east of its present course. When the Mississippi was dammed by the Grantsburg sublobe, a glacial lake was formed, and the Anoka sand plain, where the highest frequencies of Burnsi are found, is the former bed of that lake. The present distribution of both Burnsi and Kandiyohi lies almost exclusively within the recently glaciated area from which the Wisconsin ice retreated only some 10,000 years ago. Thus the period during which living things could occupy the area dates back only some 10,000 years, and the present distribution of these polymorphisms has existed for only a few thousand years. Not only are the genetics of the polymorphisms simple, but their history is brief, which should facilitate study of the adaptive role of the mutants.

Comparing the spotted wild-type pipiens with homozygous recessive albino frogs (Browder, 1967) on a grassy background brought out dramatically the cryptic value of the wild-type pattern and color. The high frequencies of Burnsi on the Anoka sand plain suggested that this unspotted mutant might confer an adaptive advantage over the spotted pipiens against a sandy background, especially in winter, when the frogs rest on the sandy bottoms of lakes and streams. It was also found that the average number of spots between the dorsolateral lines was lower in spotted pipiens frogs living within the range of Burnsi than in those living outside it, a further indication of selection for reduced spot number in this region. The fact that the progeny of low × low crosses had significantly fewer dorsolateral spots than the progeny of high × high crosses (Merrell, 1972) confirmed the heritability of spot number and showed that spot number in the spotted wild-type frogs must be influenced by modifying genes. It seems a reasonable inference that selection favors reduced spot numbers in this area in two ways, either by increasing the frequency of the Burnsi gene or by increasing the frequency of low spot number modifiers in wild-type pipiens. There was no direct evidence, however, that selection for low spot number was at work in the wild populations.

At one time, heterozygous advantage was widely regarded as the major factor responsible for maintaining polymorphisms. A major difficulty with this explanation in this case was that the available evidence showed no superiority of the Burnsi heterozygotes over the corresponding homozygotes (Merrell, 1972). Moreover, the low frequencies of the Burnsi gene posed a problem, for a simple analysis (Merrell, 1969c) showed that, to maintain the dominant at such low frequencies, the difference in fitness between the wild-type pipiens ($+/+$) and the Burnsi heterozygotes (B/$+$) could only be of the order of 0.01 or 0.001. Such a slight difference in fitness is essentially undetectable experimentally. If heterozygous advantage were in fact the explanation, one could reasonably expect the advantage to be greater and the frequency of Burnsi to be higher. Since other explanations are possible, it seems unwarranted to favor heterozygous advantage over other possibilities. Furthermore, among frogs reared in the lab where tadpoles do not thrive nearly as well as tadpoles in natural breeding ponds, there was no evidence of differences in viability among B/B, B/$+$, and $+/+$ individuals (Merrell, 1972). There was no excess of B/$+$ frogs nor any deficiency of B/B frogs, as might be expected if heterozygous advantage were involved. Similar results were obtained from the lab-reared progeny of crosses involving Kandiyohi. Therefore, the available evidence did not support heterosis as an explanation for these polymorphisms.

The low frequency of Burnsi also posed a problem for the concept that Burnsi provided better cryptic coloration than pipiens in the sandy habitat, for if that were the case, the Burnsi dominant would again be expected to have higher frequencies than those observed in wild populations. At this point serendipity crept into the picture. When Minnesota lakes open up following the spring thaw, large numbers of dead frogs and fish are sometimes found along the lake shores, a phenomenon often referred to as "winter kill." A comparison of the Burnsi frequency among the dead frogs at six different lakes with its frequency among the survivors revealed that the frequencies of Burnsi were significantly greater among the survivors than among the dead frogs (Merrell and Rodell, 1968). In lakes where samples of live frogs were taken in the fall and again the following spring, the frequency of Burnsi was higher in the spring than in the fall.

These findings suggested that frogs carrying the dominant Burnsi gene were somehow better able to survive the winter than the wild-type pipiens. The reasons for this difference are not clear, but differences in cryptic coloration seem an unlikely explanation. The difference in frequency between fall and spring live samples could be attributable to predation. However, the winter kill of frogs was accompanied by a winter kill of fish, and the major factor in the deaths of both frogs and fish appeared to be depletion of the oxygen supply in the lakes during the long winter. Dapkus (1976) also reported superior survival of Burnsi among 1,598 frogs accidentally subjected to heat stress and possible desiccation and low oxygen supply. The common denominator in those two reports is the lack of oxygen. If the Burnsi frogs are better able to tolerate low oxygen tensions, it could account for their relatively low but polymorphic frequency. Although shallow lakes are more likely to suffer O_2 depletion than deep lakes, O_2 depletion during the winter is the result of a combination of factors, including the amount of snow cover and aquatic vegetation, so that the number and distribution of lakes affected vary from year to year. Thus, selection for ability to survive low O_2 concentrations must be intermittent rather than constant, and the low frequencies of Burnsi seem in accord with its sporadic adaptive advantage. The relationship between low O_2 tolerance and Burnsi, a pigment pattern mutation, remains a puzzle. It could be a pleiotropic effect of the gene, but is more apt to be a case of linkage disequilibrium involving Burnsi and a gene or genes for low O_2 tolerance. Thus, we know a bit more than before about the possible effects of Burnsi on fitness, but each step in the investigation raises new questions.

Serendipity played a similar role in the study of the adaptive role of the Kandiyohi dominant. It had low frequencies like Burnsi but seemed to be confined to the prairie lake region of western Minnesota and the eastern Dakotas. Any cryptic advantage of its mottled spotted phenotype over the common spotted pipiens in the prairie habitat was certainly not obvious. However, some observations in the field and in the lab raised an intriguing possibility. The breeding season comes in early spring soon after the ice on the lakes begins to melt and the frogs escape from their icebound winter prison to migrate to breeding ponds. The selection of a suitable breeding pond is crucial, for *Rana pipiens* tadpoles must spend three months in the pond to complete development through metamorphosis. If the pond dries up before metamorphosis is completed in late July, all the tadpoles in the pond die. Since the average rainfall is lower on the prairie than in eastern Minnesota, the breeding ponds on the prairie are more apt to dry up than those farther east. An accelerated rate of development for tadpoles in such ponds would carry a definite selective advantage.

Therefore, when it was found that lab-reared young Kandiyohi frogs completed metamorphosis a week to ten days earlier on the average than their sibs reared in the same tank (Merrell, 1972), a possible adaptive role for Kandiyohi was suggested. This finding held true when the crosses produced Kandiyohi and pipiens siblings and also when the crosses produced sibs with four different phenotypes: Kandiyohi, Burnsi, the double recessive pipiens, and the double dominant Burnsi-Kandiyohi. The numbers of each type produced were in accord with the expected Mendelian ratios, but Kandiyohi consistently completed metamorphosis earlier than the other types of frogs. Although not all breeding ponds dry up and in some years none do, nevertheless such an event is not rare, and thus intermittent selection pressure favoring Kandiyohi must be exerted on the prairie. The fact that selection is intermittent may help to explain the low frequencies of Kandiyohi. Again the adaptive value of the mutant seems to have little connection with pigment pattern, and again either a pleiotropic effect of Kandiyohi on the rate of development or close linkage of Kandiyohi to genes enhancing the developmental rate is indicated.

The frequencies of both Burnsi and Kandiyohi seem unusually low for genetic polymorphisms. Such low frequencies present a paradox because at such frequencies, genetic drift would be expected to cause random fluctuations in gene frequency in different populations. In some, the gene would be absent; in others, present at low frequencies; and in a few, the frequencies would be quite high. Estimates of the effective size of the breeding populations (Merrell, 1968) from egg mass counts in breeding ponds showed that the effective sizes were small enough for drift to occur. However, rather than the wide range of gene frequencies expected as the result of drift, the frequencies were uniformly low and the dominants were widely distributed.

The probable explanation for this apparent paradox lies in the behavior of the frogs. The effective size of the breeding population can be better defined in *Rana pipiens* than in many species because during the brief breeding season each breeding pond contains its own distinct breeding population. Furthermore, all the tadpoles produced by these parents develop together in the same pond for three months. As soon as the parents have finished breeding, however, they disperse to their summer feeding range, and once the tadpoles have completed metamorphosis, the identity of the breeding population and its offspring as a separate entity is lost. At that time the young frogs leave the breeding ponds and migrate to the shores of larger bodies of water where they congregate by the hundreds and even thousands. At first, the small frogs stay close to water, but as they grow, they range farther and farther away in search of food. The summer movements of adult and juvenile frogs seem influenced primarily by the availability of food and moisture, and continue throughout the summer. The summer dispersal, the fall migration to overwintering sites in lakes and streams, and the spring migration to breeding ponds may involve distances of several miles. The net effect of these movements by individuals, which show no signs of group migratory behavior, is a thorough mingling of the frogs in an area and complete loss of identity of the breeding populations in different ponds. Because of the variation in precipitation from year to year and the temporary nature of the breeding ponds, different ponds are used for breeding in different years. Thus, homing to an ancestral breeding pond would be maladaptive, and there is no evidence that homing behavior helps to retain the identity of breeding populations. The wide-ranging movements of the frogs also provide the most plausible explanation for the uniformly low frequencies of the dominants. Any differences in gene frequency arising in isolated breeding ponds are quickly smoothed out by the mingling of frogs from many sources during their annual migrations and dispersals.

One intriguing question about the Burnsi and Kandiyohi dominants in Minnesota is whether all of the alleles of each mutant are identical by descent. Although the present distribution of the mutants covers some 100,000 square miles, most of this large area was occupied by glaciers up until about 10,000 years ago, so that, geologically speaking, the mutants have occupied the region only for a brief time. Given the high vagility of leopard frogs, which would permit rapid dispersal of the genes, and the brief time the polymorphisms have existed in the area, identity of descent or a common origin for all of the alleles of each mutant from a single ancestral mutation is not as improbable as it may seem at first glance. With modern molecular techniques, it should be possible to test this hypothesis.

This brief recital of an attempt to study the adaptive role of two dominant genes in natural populations of a common species raises more questions than it answers, but it does illustrate some of the problems associated with the study of adaptation. Armchair speculation about the adaptive significance of a gene or a trait is easier than a study in the field and has provided endless diversion for those so inclined. The conclusions drawn from studies by means of computer simulation flow directly from the data and assumptions used to set up the program. Because simplifying assumptions are generally necessary, simplistic conclusions are almost inevitable. Study of the autecology of a species requires total immersion of the observer in the life history of the species and is the first prerequisite for a study of the adaptation of a species to its physical and biological environment. What emerges is an appreciation of the complexities of adaptation, survival, and reproduction in the lives of even the simplest of organisms. Solutions to adaptive puzzles do not necessarily fall out of such studies like ripe plums, but the more that is learned about a species, the easier it becomes to formulate plausible, testable hypotheses about the adaptive significance of the observed features of the species and its life history. They also lead to a realization that adaptation is an ongoing, pluralistic process, never perfected because the biological and physical environment is constantly changing. For this reason the adaptive paradigm is best visualized as an adaptive seascape.

Chapter 11
Controversies about Evolution

In this chapter, we shall deal with some of the controversies that have arisen during the study of evolution. In many cases, these controversies have developed because evolutionists have taken a dialectical approach to their studies, advocating one or the other of alternative theories. However, as Simpson (1983) pointed out, "One should always be wary of an 'either-or' proposition." Because living and extinct species are so numerous and so diverse, it seems highly unlikely that they have all evolved in the same way. Rather than trying to force them all into a single theoretical framework, it seems preferable to recognize that theoretical pluralism will provide the most complete and reasonable interpretations of the data.

Approaches to Evolution

In part, the differences of opinion among evolutionary biologists stem from their differences in background and in their approach to evolution. For example, many years ago Simpson (1944) wrote: "Not long ago paleontologists felt that a geneticist was a person who shut himself in a room, pulled down the shades, watched small flies disporting themselves in milk bottles, and thought that he was studying nature. . . . On the other hand, the geneticists said that paleontology had no further contributions to make to biology, that its only point had been the completed demonstration of the truth of evolution, and that it was a subject too purely descriptive to merit the name 'science.' The paleontologist, they believed, is like a man who undertakes to study the principles of the internal combustion engine by standing on a street corner and watching the motor cars whiz by." In view of these comments, it is worth noting that the most quoted recent attack on the adaptive paradigm was launched by a paleontologist and a *Drosophila* geneticist (Gould and Lewontin, 1979). Perhaps Simpson's assessment is still valid, for neither a paleontologist nor a *Drosophila* geneticist can easily evaluate or appreciate the subtleties of adaptation to natural conditions in the species they study. Because of the nature of the fossil record, paleontologists are limited in their capacity to study the ecology and adaptation of extinct species, and most *Drosophila* geneticists still study flies disporting themselves in bottles rather than in nature.

Among neontologists, those using a reductionist approach (e.g., the molecular biologists, biochemists, cell biologists, and most of the physiologists) are unlikely to develop much insight into the adaptation of organisms to their environments because adaptation and adaptability are properties of whole organisms. Those using a holistic approach, who study the emergent properties of organisms, have a better opportunity to understand such relationships. Reductionists can usually test their hypotheses experimentally, however, whereas holistic biologists too often must rely on observation and the comparative method rather than experiments to test their theories.

Another dichotomy in the approach to evolution is that between the experimental naturalists and the mathematical theorists (Waddington, 1953). The experimental naturalists, now called ecological geneticists, work to resolve evolutionary questions through a combination of field and laboratory studies. The mathematical theorists work from first principles based on Mendelian genetics, trying to predict the course of evolutionary change. With so many evolutionary factors at work and with so many genes involved, the mathematics quickly becomes intractable and simplifying assumptions become necessary.

Even among the mathematical theorists, there are two schools of thought, population genetics and quantitative genetics, which, as noted earlier, have surprisingly little in common. Population geneticists analyze the possible effects of mutation, selection, migration, and drift at one or a few loci, and, by extrapolation to many loci, try to reach generalizations about the evolutionary process. Quantitative geneticists start at the other end of the spectrum with continuous phenotypic variation and attempt to reconcile particulate Mendelian genetics, which leads to discontinuous variation, with the observed variation in quantitative traits. This reconciliation has been achieved by assuming that continuous phenotypic variation results from the combined action of many genes, each of small effect, at a number of loci, plus the effects of the environment, and takes into account the effects of dominance and epistasis and of genotypic × environmental interactions. Where the population geneticist deals with gene and genotype frequencies in natural populations, the quantitative geneticist deals with variances, covariances, and heritabilities in populations of domesticated plants and animals. Applied to plant and animal breeding, quantitative genetics has been enormously successful, but it has not been equally successful in evolutionary biology.

Quantitative genetics is also known as statistical or biometrical genetics, which is the mark of its greatest strength and its greatest weakness, for the statistical sophistication of quantitative genetics is not matched by a comparable genetic sophistication. An understanding of the mechanism of evolution requires an understanding of the genetic mechanism underlying evolution that is at least as sophisticated as the statistical aspects of quantitative genetics and is not based simply on the concept of additive genetic variance. The preoccupation with the phenotype in quantitative genetics and the numerous simplifying genetic assumptions sometimes leave the impression that one is dealing with voodoo genetics.

One of the earliest either-or controversies about the mechanism of evolution was that between the selectionists and the mutationists early in this century. De Vries's mutation theory was regarded not as complementary to Darwin's theory of natural selection but as an alternative to it. This conflict was resolved by the work of the population geneticists, who showed the essential and complementary roles played by both natural selection and mutation.

Another early conflict lay between those who believed that evolution was always gradual and mediated by polygenes and those who believed that evolution could occur in steps mediated by oligogenes. In some ways, this is another way of stating the conflict between the selectionists and the mutationists, for the selectionists were usually gradualists and the mutationists were usually saltationists. The gradualist position has become dominant in present-day evolutionary thought despite compelling evidence for an evolutionary role for oligogenes and chromosomal rearrangements. Among the leading theorists, only Sewall Wright (e.g., 1949) continued to include these factors within his theoretical framework, but he seemed to do so more for the sake of completeness than because of any strong belief in their importance. His shifting balance theory is a theory of gradual evolution.

The most bitter and protracted controversy about evolution arose between Fisher and Wright over the evolution of dominance. Neither seemed able to recognize much merit in the other's position, yet, as we saw earlier, the origin of dominance may involve elements of several theories because dominance may originate in different ways. Rather than a rigid either-or, right-or-wrong dichotomy, what is needed to explain the origin of dominance is a theoretical and explanatory pluralism. Again the dispute can be framed in terms of polygenes versus oligogenes, for Fisher's theory involved numerous modifier genes of small effect whereas Wright's theory was focused on the action of genes at particular loci.

Another controversy arose over whether population genetics had made significant contributions to the study of evolution. Waddington (1953), for example, wrote, "It has not . . . led to any noteworthy quantitative statements about evolution"; and, "Very few qualitatively new ideas have emerged from it." Mayr (1959) somewhat disdainfully referred to population genetics as "beanbag genetics" because he felt that population geneticists treated genes in populations as independent units like beans in a bag and, ignoring genic interactions, regarded evolution as simply a change in the proportions of the different kinds of beans. The mathematical constraints that forced the early population geneticists to concentrate on a single locus may have made it appear that they were ignoring the effects of genic interaction, but that was far from the case. Although Fisher was already dead by that time, Wright (1960) and Haldane (1964) both responded to these attacks, Haldane by defending "beanbag genetics" and Wright by pointing out that factor interactions played an integral role in the theories of evolution both he and Fisher had developed. And it is true, of course, that Wright's shifting balance theory and Fisher's theories of selection and the origin of dominance all involve various kinds of genic interaction. An appreciation of the evolutionary importance of genic interaction was present from the early days of population genetics and not just after 1940, as Mayr maintained.

Classical and Balance Theories of Variation

Another long-standing dispute in evolutionary biology developed over the nature of genetic variation in sexually reproducing wild populations. The "classical" theory, most closely identified with Muller (1950), postulated that each individual in a population is homozygous at nearly all loci for the "wild-type" allele at that locus, with a small number of loci heterozygous for rare deleterious genes. The "balance" theory, usually associated with Dobzhansky (1955), posited that each individual in a population is heterozygous at nearly all loci, with homozygosity primarily resulting from matings between close relatives. Under the balance theory, wild-type or "normal" alleles cannot be identified because all the "normal" individuals in a population are heterozygotes. This dispute developed because of the difficulties in assessing the amount of genetic variation in wild populations. As long as the studies dealt with visible mutants or lethals, it was impossible to tell which theory was closer to reality. When analysis of electrophoretic variation in enzymes and soluble proteins began in the 1960s (Hubby and Lewontin, 1966; Lewontin and Hubby, 1966), the discovery that about a third of all loci studied were polymorphic and that the average proportion of heterozygous loci per individual was about 10 percent provided strong support for the balance hypothesis, and the controversy seemed to be resolved.

The electrophoretic studies proved to be too much of a good thing, however. Although less than half of all the genetic variation was being detected by electrophoresis, more variation was found than could be comfortably explained by most of the fashionable theories of the day. In particular, heterosis was the most popular explanation for genetic polymorphism at that time, and the adaptive advantage of the heterozygotes over the corresponding homozygotes was thought to be responsible for the persistence of allelic variation in populations. Heterozygote advantage also generates a genetic load, however, because the less fit homozygotes are constantly being produced by segregation in the heterozygotes. As the number of polymorphic loci increases, Lewontin and Hubby calculated that the segregation genetic load would become intolerably great, that all individuals would be homozygous for at least one deleterious gene and thus would suffer from reduced fitness. This conclusion seemed to pose an embarrassing dilemma for advocates of heterosis.

One effort to salvage the heterosis theory involved truncation selection (Sved, Reed, and Bodmer, 1967; King, 1967; Milkman, 1967). This type of selection assumes a fitness threshold, above which an individual survives and reproduces and below which it does not. If the probability of exceeding the fitness threshold increases with increasing heterozygosity, then selection will tend to eliminate individuals carrying an excess of deleterious genes in the homozygous condition, and each genetic death will cause the elimination of a number of harmful genes in a single selective event. If an upper asymptote for fitness is assumed as well as a selective threshold, the heterotic explanation for genetic polymorphisms becomes more credible. Nevertheless, one flaw remained, for the expected amount of inbreeding depression with the heterotic model was much greater than the amount actually observed. (For further discussion of genetic load, see Wills [1981], Milkman [1982], and Wallace [1991].)

With heterosis on shaky ground as the explanation for the great amounts of genetic variation being revealed, geneticists turned to other forms of balancing selection for an explanation. These other theoretical possibilities included frequency-dependent selection, density-related selection, and heterogeneity of selection pressures in time or in space. Thus, a theoretical pluralism emerged, but these theories for the most part remained speculative because of the usual difficulties in demonstrating the workings of natural selection and adaptation in wild populations.

Adaptive Neutrality

Another major effort to explain the high levels of genetic polymorphism in natural populations took an entirely different tack. Rather than explanations based on some form of balancing selection, Kimura (1968; Kimura and Ohta, 1971) and King and Jukes (1969) asserted that the alternative alleles were adaptively neutral. This neutral theory of molecular evolution, set forth in detail in Kimura (1983), is also known as the neutral mutation-random drift theory or non-Darwinian evolution or simply the neutralist theory because it is a theory of evolution that minimizes the role of natural selection. Many evolutionary biologists found neutralism difficult to swallow and vigorously opposed it. The controversy between the neutralists and the selectionists dragged on for years, primarily because neither side could come up with decisive evidence for its position. Eventually, after playing a dominant role in population genetics and evolutionary biology for some twenty years, the controversy seemed to fade away, with neither side able to claim victory or willing to concede defeat, and an undeclared truce seems to have set in.

Although still unsettled, the controversy between the selectionists and the neutralists has been instructive for several reasons. First, the efforts to resolve the controversy stimulated a great deal of worthwhile research. Further, it pointed up the hazards of the either-or approach to evolutionary questions, for the positions taken by the neutralists and the selectionists seemed far too inflexible, leaving little room for compromise. Lewontin (1974) held that the neutralist theory was simply a revival of the classical theory in modern dress and labeled it the neoclassical theory. The trouble with the neutralists' position was that they seemed to have a theoretical answer to every objection but could never provide conclusive proof of their beliefs. They always had fallback positions ready whenever a particular position became untenable. The selectionists, however, also failed to prove their theories, which remained articles of faith rather than experimentally established facts. The attack mounted by Gould and Lewontin (1979) on the "adaptationist programme" was made possible—indeed, was almost invited—by the propensity of the selectionists to speculate about adaptation and selection rather than to conduct actual tests of their hypotheses. In this case, the hardening of the positions of the neutralists and the selectionists seems to have blinded both sides to the merits of the others' position. Evolution is a combination of random and nonrandom events, and the biochemical variation revealed by electrophoresis seems no different from other forms of variation in being molded by a combination of chance and selective events. (Kimura's 1983 book should be consulted for an authoritative exposition of the neutral theory of molecular evolution.)

Characterizing abundant electrophoretic variation as selectively neutral is appealing because if this variation is neutral, no function need be sought. Similarly, describing the noncoding or untranscribed portions of the genome as genetic "junk" or "selfish DNA" categorizes it as unimportant to the life of the organism. The fact that structural DNA coding for proteins has an obvious function, but no function is immediately apparent for "neutral" alleles or for "junk" DNA, hardly constitutes proof that such DNA lacks a function. In view of how little the operation of the genetic system in the development and adaptation of organisms is understood and how complex the developmental history and the life history of organisms can be, at present it seems advisable to keep one's mind open to the possibility that the subtleties of differentiation and adaptation are mediated, not just by the structural DNA, but also by the untranscribed regions of the DNA.

The idea that the noncoding DNA may consist of regulatory genes and that these genes may play a significant evolutionary role is certainly not new (e.g., Wilson, 1975). However, the concept of regulatory genes is rather amorphous and at the same time somewhat restrictive. Structural genes may also have regulatory functions, so that the distinction between regulatory genes and structural genes is ambiguous. Furthermore, the idea that regulatory genes simply activate or inactivate structural genes may give a limited view of the actual diversity of function that resides in the noncoding part of the genome. The idea that self-perpetuating junk DNA can parasitize the genetic system of evolving organisms has the appeal of "explaining away"

(Cain, 1979) its presence, but leaves one with an uneasy urge to know more about how the genetic system really works before coming to such a far-reaching conclusion.

Because only a fraction of the structural genes are active at any given time, most structural genes will appear to be inactive or neutral most of the time. If the rest of the DNA—the regulatory genes, the selfish genes, the genetic junk, the pseudogenes, the introns, the parasitic DNA, the repetitive DNA, and other nonspecific DNAs—also function intermittently, they too may appear to be neutral. Given this situation, it is not surprising that so little has been learned about the functions of this residual DNA.

Furthermore, even though both structural and nonstructural DNA may be neutral much of the time, it still may affect fitness. A favorite assumption of the neutralists, for example, is that synonymous codons are adaptively equivalent because they code for the same amino acid. This assumption needs to be tested, however, for if a different tRNA is involved or if the protein is produced at a different rate, there may be significant fitness differences between the so-called neutral alleles despite the fact that they code for the same amino acid.

If, instead of assuming there is so much genetic variability in natural populations that it cannot all be functional, we look at the other side of the coin and try to envision what this DNA is called on to do, we may gain a somewhat different perspective. In essence, it guides the differentiation, development, and adaptation of an organism throughout its life cycle from fertilized egg through maturity and affects traits ranging from the molecular to the behavioral. In other words, the DNA in the genetic system does far more than simply code for proteins.

The genotype can be thought of as functioning in a multidimensional system, the three dimensions of the developing organism itself plus the many dimensions of the environment, all changing through time. Development is not simply the unfolding of a preprogrammed plan of differentiation encoded in the genotype, for developmental flexibility or developmental homeostasis or both may be manifested, depending on the environmental conditions to which the developing organism is exposed. These types of adaptability are under genetic control, as are various kinds of short-term physiological adaptation. The fine-tuning of the developing organism, both in its internal coordination and its physiological adaptations to changing external conditions, is clearly influenced by heredity.

Adaptation and Molecular Evolution

These capabilities are embedded in the genotype, and the question is where. If we think, not in terms of a fixed phenotype, a fixed fitness, and a fixed environment, as has so often been the case, but instead in terms of an adaptive seascape to which organisms must constantly adjust, we may search more diligently for the adaptive significance of the large amounts of genetic polymorphism and of the seemingly useless DNA that now is regarded as cluttering up the genome.

At present, understanding of the mechanisms by which the genotype influences development, adaptation, and behavior is in its infancy, but at least the importance of heredity in these processes is finally being recognized (Nevo, Beiles, and Ben Shlomo, 1984; Nevo, 1988, 1991). It seems unlikely that the structural DNA coding for proteins provides all of the genetic nuances affecting these processes, which means that we must look elsewhere in the genome. Elsewhere means another look at "neutral" alleles and all the untranscribed portions of the genome to which no function has yet been ascribed. That an organism as efficiently functioning and highly adapted to its environment as, say, a leopard frog in Minnesota does so by means of a group of structural genes grinding out proteins plus an array of neutral alleles and an assortment of selfish, parasitic, or junk DNA seems a bit much.

That natural selection should operate so effectively on the development, adaptation, and behavior of organisms, but not on the molecular processes undergirding these phenomena, seems hardly credible. Worse yet, that these hereditary effects should be mediated by a conglomerate of genetic junk unscreened by natural selection seems equally improbable. The pathways between the DNA and its phenotypic effects on development, adaptation, and behavior may be long and complex and difficult to trace, but there is substantial evidence that they exist. If that is so, then when natural selection affects the developing, adapting, or behaving organism, its effects should ultimately be reflected by changes in the genetic system at the molecular level. In other words, molecular evolution and morphological, developmental, physiological, adaptive, and behavioral evolution must in some way be interconnected. To assume that two different, independent modes of evolution exist, one largely random for molecular evolution, the other highly selective for evolution at higher levels, seems to fly in the face of reason and the evidence.

Molecular and Morphological Evolution

Nevertheless, a controversy has arisen over the relationship between morphological and molecular evolution. King and Wilson (1975), for example, concluded that molecular and "organismal" evolution must occur independently because they found the average human polypeptide to be more than 99 percent identical to its chimpanzee counterpart, giving a genetic distance comparable to that of sibling species, yet chimpanzees and humans "differ far more than sibling species in anatomy and way of life." In fact, taxonomists classify humans and chimpanzees not only in separate genera but in separate families, Hominidae and Pongidae, because they consider them to be so different.

That morphological and molecular evolution might sometimes occur independently is hardly surprising, for different molecules may also evolve independently of one another and so may different morphological traits. Even different parts of the same molecule may evolve independently. However, because the observed differences in "anatomy and way of life" were not reflected in the macromolecules King and Wilson studied, they concluded that the differences between chimpanzees and humans must be the result of differences in regulatory genes rather than in the genes coding for polypeptides.

This intriguing conclusion invites scrutiny, for it is a transcendent leap from the springboard of the data. It is difficult to know where to begin. Taxonomy, as it gets closer to humans, seems to fall apart; new species, new genera, and new families are named at the drop of a hat. The criteria normally applied to other animal groups are tossed out the window, and a new name is attached to every fossil hominoid bone that turns up. Anthropocentrism is undoubtedly at the root of this proliferation of names. Because taxonomic names are the subjective creations of taxonomists and taxonomists are human, this separatist tendency is not particularly surprising.

To the somewhat irreverent suggestion (Merrell, 1975) that perhaps, in light of their molecular similarities, humans and chimpanzees might not be so different after all, a spirited defense of the uniqueness of humans was mounted by a biochemist and his associates (Cherry, Case, and Wilson, 1978). Oddly enough, they seemed intent on proving that the morphological evidence on the relationship between humans and chimps is more compelling than the biochemical evidence. However, their choice of morphological traits for comparison was seriously flawed. In an effort to avoid bias, they chose nine skeletal traits previously used to distinguish among separate populations of tree frogs, but they totally ignored the fact that frogs, which are all amphibious species to a greater or lesser degree, have similar body shapes because they have similar modes of locomotion. Chimpanzees, on the other hand, are an arboreal species, moving by brachiation, whereas humans are terrestrial, with erect bipedal loco- motion. Because they are adapted to different modes of locomotion, their skeletal measurements have diverged markedly. Thus, Cherry et al. could hardly have chosen a less reliable set of morphological traits on which to base their comparisons.

Unfortunately, very little is known about the genetics of the skele- tal traits used in these comparisons. The human and chimpanzee bones themselves are much alike and are clearly homologous. Where they differ is in their relative proportions. If the differences between chimpanzee and human skeletons result from differences in allomet- ric growth (Huxley, 1932), the genetic changes involved could be rel- atively simple.

A similar case was reported by Carson (1970) for the Hawaiian spe- cies, *Drosophila silvestris* and *D. heteroneura*. These two species show a number of morphological differences, the most striking being the un- usual mallet-shaped head of *heteroneura*. Thus, they are not like sib- ling species, which are difficult to separate morphologically. None- theless, allozyme comparisons showed that the enzymes of the two species are as similar as those normally found in different local pop- ulations of the same species. After taking all of the other evidence into account, Carson decided that these two species were of recent origin and were very closely related despite their striking morphological dif- ferences. Wilson and his associates might well have reached a similar conclusion about humans and chimpanzees if humans had not been one of the species involved. Faced with an apparent contradiction be- tween the morphological and the biochemical data, they felt con- strained to rely on one or the other. The fallacy of the either-or ap- proach is clearly illustrated in this case, however, for the total available evidence indicates that humans and chimpanzees are not as distantly related as Wilson's nine morphological traits suggest nor as closely related as the biochemical data indicate.

The biochemical data also raise questions. King and Wilson (1975) based their original comparison of humans and chimpanzees on forty-four loci. As such studies go, that is quite a few, but compared to the estimated thousands of structural genes in these species, it does not seem like so many. Furthermore, a selected sample of loci was used, so that perhaps it is not surprising that few differences were found. From this limited sample of loci, they leaped to the conclusion that the species differences resulted from the action of regulatory genes rather than structural genes. That may very well be true, but it is not very strongly supported by the failure to find significant differences in such a sketchy sample of structural gene loci. Subsequently, Wilson and his colleagues (Wilson et al., 1975; Larson, Prager, and Wilson, 1984) also stressed the importance of differences in chromosomal structure and changes in social behavior to the formation of species. All these suggestions, including the role of regulatory genes, seem worth pursuing, but at present there is only a modicum of evidence to support such wide-ranging speculations.

Cultural Evolution

An interesting aspect of the work by this group is that they have revived the nature-nurture controversy in a new guise, another instance of the hazards of the either-or approach in biology. Sage et al. (1984) wrote, "It has been suggested that cultural drive, a potent nongenetic force, may be responsible for the rapid morphological evolution and speciation in some mammal and bird groups." By cultural drive, they apparently meant rapid evolution by means of behavioral innovation and social learning. However, the idea that cultural drive is a nongenetic force is clearly erroneous. Just as there is always a genetic and an environmental component in the development of any morphological trait, there is also a genetic and an environmental component in the development of any behavioral trait. The acquisition of language by human infants is a case in point. The particular language learned is a function of the cultural environment in which the infant is reared, but the ability to learn human speech is clearly genetic. Despite considerable exposure to humans and even deliberate efforts to teach them to talk like humans, no other primate has ever mastered a human language. They simply lack the genetic capacity to learn to speak. Therefore, to think of cultural evolution, behavioral innovation, and social learning as completely divorced from heredity represents a misconception of how the evolutionary process works. Even learning and cultural evolution are colored and shaped by the genotypes of the organisms involved. The question, as always, is not nature or nurture, but how do nature and nurture interact to produce the observed phenotypes.

Evolutionary Complexity

Kolata (1975), commenting on the King and Wilson (1975) paper, wrote, "Wilson and his associates report that frogs, which are anatomically simple organisms, exhibit fewer chromosomal changes than mammals, which, of course, are more complex." Again anthropocentrism seems to have reared its ugly head. The appropriate question really is, are mammals more complex than other groups, or better yet, in what ways are mammals more complex? If metabolic versatility is the criterion, bacteria are metabolically much more capable than mammals, for they can synthesize virtually all their essential metabolites from simple inorganic compounds, something mammals cannot do.

As for anatomical complexity, frogs are at least as complex as mammals, and, if all stages of the life cycle are considered, frogs clearly emerge as the winners, for their life cycle is more complicated than that of mammals. Frogs have long been dissected in beginning biology labs as a representative vertebrate, and the homologies in the anatomy of adult frogs and mammals are unmistakable, so that as adults they seem equally complex.

In addition to the adult stage, however, the frog life cycle includes a free-living aquatic tadpole stage absent in mammals. The genetic system that controls the differentiation of the early embryo into an aquatic, gill-breathing, herbivorous tadpole and then the metamorphosis of the tadpole into a terrestrial, lung-breathing, carnivorous tetrapod has to govern much more than just the resorption of the tail and the differentiation of legs and lungs. The tadpole excretes nitrogenous wastes as ammonia; the adult frog excretes urea. The visual pigment in tadpoles is porphyropsin; in the adults it is rhodopsin. Even the adult hemoglobin differs from the tadpole's in having a decreased affinity for oxygen and in showing the Bohr effect. Thus, the tadpole, with structural and biochemical adaptations like those of freshwater fish, metamorphoses into a frog with structural and biochemical adaptations characteristic of terrestrial tetrapods. Even the adult frog must be capable of adapting to two very different modes of life, solely aquatic in the winter and amphibious the rest of the year. Its genotype must be at least as versatile as that of mammals to meet the variety of demands placed upon it. Therefore, even though Kolata suggested that more chromosomal changes occurred during the evolution of mammals than during that of frogs because mammals are more complex, this concept fails to hold up under close scrutiny because in many respects frogs are more complex than mammals. It appears that the explanation for the greater numbers of chromosomal changes in the evolution of mammals compared to frogs must be sought elsewhere.

The relationship between natural selection and sexual selection is another area where a continuing difference of opinion exists. Darwin regarded sexual selection as an evolutionary force independent of natural selection, a view still held by many biologists (e.g., O'Donald, 1980; Tomlinson, 1988) who argue that sexual selection sometimes opposes natural selection. Population geneticists, on the other hand, tend to regard sexual selection as just one of the many components of fitness under the rubric of natural selection. This disagreement is traceable to a difference in interpretation of the meaning of fitness. For instance, fitness is often measured in terms of and equated with viability or fecundity or longevity. If this is done, sexual selection can be regarded as an independent factor in evolution. As noted earlier, however, fitness is best defined in terms of the numbers of progeny left by an individual or in terms of its contribution to the gene pool of the next generation. Fitness has a dual nature, for it entails both the survival and the reproduction of the individual. Under this definition, sexual selection becomes a component of fitness, equivalent to viability, fecundity, and longevity, rather than independent of or even opposed to natural selection.

Group Selection

Another debate has raged over the role of group selection as distinct from individual selection. Again the dispute stems in part from the way the concepts are defined. The modern discussion of group selection originated with the work of Wynne-Edwards (1962), who proposed that through group selection a population evolves restraints to its own reproduction so that it will not overtax its resources. The concept of group selection was subsequently broadened during attempts to explain other group adaptations such as cooperative behavior, warning signals, schooling behavior, altruism, insect societies, and the sex ratio. The semantic entanglements that arose were formidable (Williams, 1966; Wilson, 1983), for different authors had somewhat different concepts of group selection. We shall not even attempt to pursue all the ramifications of the subject.

Although some authors have attempted to explain all supposed cases of group selection in terms of individual selection, at present it seems desirable to leave both options open, to favor explanatory pluralism rather than an either-or approach. When all facets of the subject are considered, however, group selection does not loom very large in comparison to individual selection as a mover and shaker on the evolutionary scene.

If we assume that individual selection and group selection are separate and independent evolutionary factors, then both individual and group selection might act in the same direction. In such a case it would be difficult to distinguish between them, but evolutionary change might be greater than if either were acting alone. On the other hand, if individual selection and group selection are in opposition, the distinction between them should be easier to make.

In an earlier chapter, we discussed individual phenotypic traits and also group traits such as population density, effective size of breeding population, migration rate, birth and death rates, and biotic potential. Group selection has apparently not been used to account for the evolution of group traits such as these, which have usually been attributed to the action of individual selection. Instead, group selection has been invoked in attempts to explain the evolution of traits that seem to run contrary to the interests of the individual, but favor the group as a whole.

As noted earlier, these traits include such things as altruism, cooperative behavior, and warning signals. For example, a warning signal by a bird alerts other members of the flock to the presence of a hawk but betrays its own position to the predator. Selection theory suggests that the bird would enhance its chances of survival by remaining silent. Thus the evolution of warning signals seems to run counter to expectations based on individual selection, and an explanation based on group selection was proposed. It was postulated that, despite the greater risk involved for the individual giving the warning signal, this altruistic behavior had evolved because it enhanced the chances of survival for the group as a whole. The difficulty with this proposal is the little matter of the lack of benefit to the "brave and virtuous" bird that gives a warning to others. If birds tending to give warning calls run a greater risk of predation than those remaining silent, and if this tendency is to any extent hereditary, the flock will soon consist only of birds that remain silent when hawks are about. Group selection theory often got hung up on dilemmas like this, but the fact remained that warning calls had evolved.

An escape from the dilemma emerged with the theory of kin selection (Hamilton, 1964). Kin selection was somewhat disillusioning, for if true, it meant that the altruistic bird giving the warning signal was acting in its own self-interest after all. The reasoning is that the other members of the flock are apt to be its relatives, and in giving them notice of approaching danger, the bird is behaving in such a way as to increase the chances that the genes it shares with its relatives will be preserved even though its own genes may be lost in the process. So much for pure altruism!

The concepts of the extended phenotype and inclusive fitness (Hamilton, 1964) were brought into play here, the idea being that the phenotype of an individual is not delimited by its ectoderm nor is fitness simply a function of the phenotype thus narrowly defined. The concept of inclusive fitness is usually presented as quite novel, but long before it was introduced by Hamilton, animal breeders were using family selection and sib selection in their work. For example, in dairy cattle they used the milk production of sisters to estimate the breeding value of a bull, and for traits with low heritability, they used family selection rather than individual selection as a means to make more reliable genetic gains. Thus, the essentials of the concept of inclusive fitness were embodied long ago in the work of the animal breeders.

Thoughts about the evolution of altruism were carried a step further by the concept of reciprocal altruism (Trivers, 1971). In this case it was postulated that one individual would help another, even an unrelated individual, if it expected the other to reciprocate at a later time. In this case, there would be no need for a close genetic relationship, for the mutually beneficial behavior would help both individuals to survive and reproduce. If this scenario is true, then the back-scratching behavior of politicians is mimicked in nature, and self-interest rather than pure altruism again prevails. What is more, this sort of behavior is explicable in terms of individual selection rather than group selection.

The upshot of this work is that altruism and various kinds of cooperative behavior as well as such group characteristics as birth and death rates, population density, and migration rates are readily explained as the products of individual selection. When the survival and reproduction of individuals are promoted by cooperative behavior rather than competition, the genes favoring cooperation in these individuals will increase in frequency in the population, and the result is a straightforward example of individual selection. Kin selection and reciprocal altruism help to account for some seemingly anomalous cases of behavior in or by groups, but the basis for both kin selection and reciprocal altruism is individual selection, not group selection.

Nevertheless, the group selectionists have persisted in their quest for group selection and cite population selection, species selection, and even community and ecosystem selection as examples of group selection (Lewontin, 1970; Wilson, 1980; Sober, 1984; Brandon and

Burian, 1984). The question remains as to whether any plausible cases of selection can be identified at these higher levels. The whole concept is tied to the belief that there is a hierarchy of levels of selection, that the units of selection range from the molecular and the cellular through the individual to the higher levels of population, species, and ecosystem. As stressed earlier, however, individual phenotypic selection is the level at which natural selection operates. Genes and molecules, for example, are not selected for and by themselves, but for their effects on the phenotypes of individuals in a breeding population. Nonetheless, some (e.g., Ghiselin, 1974; Hull, 1976, 1978) have argued that a population or a species or even a community has the characteristics of an individual or superorganism. In this way selection at these higher levels has been reduced to instances of individual selection, an exercise in semantics.

The greatest difficulty with group selection is to find a credible mechanism by which it can operate. Before considering species selection or selection at even higher levels, let us consider population selection or selection between different demes of the same species. In this case a plausible mechanism involves a combination of interdemic selection and gene flow. Species usually consist of a number of widely distributed and relatively isolated demes, each pursuing its own more or less independent evolutionary path. If, through a combination of favorable genetic and environmental circumstances, one deme becomes better adapted to the existing conditions than others, this change will be reflected by an increase in the numbers of this deme. The pressure of numbers will cause members of the deme to disperse, introducing their favorable genes into other demes by gene flow. In this way, genes from the more successful deme can eventually spread throughout the species. This explanation, based on concepts first enunciated by Sewall Wright, has the merit of providing a readily understood mechanism for interdemic selection, but it may be too mundane for some and may not even meet the definition of group selection for others. In such a situation, the superiority of the deme is based on the superiority of the individuals of which it is composed rather than on the superiority of one deme over another. And this seems to be true in most instances where group selection is postulated; the merits of the group, whatever they may be, are based to a far greater degree on the merits of the individuals in the deme than on the merits of the group as such versus other groups.

Now let us consider species selection. The usual concept is that when two closely related but ecologically different species compete with one another, the more successful one will preempt the available space and resources and drive the other to extinction. The main trouble with this picture is that things seldom work out this way. If species are competing, there should be some evidence for this competition. The available evidence indicates that intraspecific competition between members of the same species is far more severe than competition between members of different species. Related species often exist sympatrically without competition or conflict because their ecological niches differ, an avenue of conflict avoidance not ordinarily available to members of the same species.

The usual explanation for these niche differences is that disruptive or diversifying selection has resulted in character displacement in the "competing" species, thus minimizing the competition between them. The evidence on the matter is a mixed bag, however, for comparisons of species pairs in areas of sympatry and in areas of allopatry sometimes reveal greater differences in areas of sympatry, as expected under the theory, but in other cases the two species are more alike in the area of overlap than when allopatric, and in still others sympatry is without effect (Mayr, 1963). Convergence in areas of sympatry can be readily explained as the result of similar responses to a common selection pressure while the same degree of similarity between species no matter whether sympatric or allopatric can be regarded as evidence that they are not competing at all. However, these results hardly provide rigorous support for character displacement, and where everything can be explained so easily, probably nothing is explained, for the explanations are too facile to make them totally convincing.

The importance of interspecific competition is still a significant question in evolutionary biology, but whether species win or lose in the struggle to persist by species selection as such is still a moot question. The introduction of starlings into North America led to a notable decline in the bluebird population because starlings easily evicted bluebirds from their nesting holes. This was certainly a case of competition between two species, but the outcome depended on the results of a series of individual encounters over nesting sites in which the bluebirds almost invariably lost out to the starlings. Thus it is hardly an unambiguous case of species selection.

The fate of the passenger pigeon may also be instructive, for the passenger pigeon is extinct—as dead, it can be said, as the dodo. Various theories have been proposed to account for its extinction, but no one has yet called it the result of species selection. The decline of the American chestnut from chestnut blight and that of the American elm from Dutch elm disease have virtually eliminated these species from the eastern North American hardwood forests but can hardly be attributed to species selection unless a very different twist is given to the meaning of species selection. The elm and the chestnut have succumbed to two species of fungi because they lacked resistance to them. Again one species has won out over another, but the host-pathogen relationship is not what is ordinarily meant by species selection.

The examples of the passenger pigeon, the American chestnut, and the American elm are instructive in another way. All were major elements in the eastern hardwoods forest, a community, a biome, or an ecosystem, as you wish. Each of these species has been or is being eliminated from the community, but the community continues to persist. Although it may be helpful to talk in terms of community or biome or ecosystem selection for didactic purposes, selection at these levels seems highly improbable. Changes in communities are determined by the evolutionary changes in their component species, and the changes in the component species occur primarily as the result of the survival and reproduction of the individual members of each species in the face of the vagaries of their day-to-day existence. The difficulty with species selection and community selection is that the concepts are nebulous, no mechanism has been suggested by which they could occur, and no examples have been provided for study. The mere fact that when a group of related species appears in the fossil record, some species persist while others disappear, hardly constitutes evidence for species selection. Similarly, if some faunal groups of diverse species persist for millions of years in the fossil record while other faunas come and go, community selection is not the most probable explanation for these observations. Until some plausible mechanism is proposed to explain how species selection or community selection or ecosystem selection can operate effectively and independently of individual selection, a certain degree of skepticism about the significance of selection at these levels seems warranted.

The Origin of Species

Next we shall consider briefly some of the controversies associated with the origin of species, which again have so often been cast in the form of alternatives, with the adherents seeming to favor one or the other. Although Mayr (1988) has recently argued in favor of explanatory pluralism in evolutionary biology, he has been one of the more vigorous proponents of a single explanation for the origin of species. Specifically, he has long maintained that speciation is allopatric and has argued at length against the importance of sympatric speciation. However, such a position is hard to sustain in view of the large numbers of allopolyploid plant species and the recent work (Tauber and Tauber, 1989) on sympatric speciation in animals. There can be little doubt that both sympatric and allopatric speciation have occurred.

Another controversy has been very one-sided, for the great majority of evolutionists believe that evolution has been gradual, and the few who, like Goldschmidt, have supported saltational evolution have often been ridiculed. A corollary to this is the belief that gradual evolution has been mediated by polygenes governing quantitative traits (Carson, 1982). Allopolyploidy again rears its ugly head, however, for it is about as saltational as speciation can be. Moreover, now that karyotyping is more easily done, it has become clear that speciation in most plant and animal groups has been accompanied by chromosomal rearrangements (White, 1978), and allopolyploidy can no longer be regarded as an aberration from the usual mode of speciation. Speciation may be genic, *a la* Carson, but it may also be chromosomal, and most of the time it is probably both. It may be gradual, but it may also be quite abrupt.

The value of the debates and the controversies about evolution is the stimulus to research they have provided. Ideally, experiments are designed to support or falsify a hypothesis, to provide a clear-cut answer as to whether a concept is right or wrong. Perhaps the way scientists' minds work is dictated by the nature of their approach to problems. They are eager to seek a single explanation, to use Ockham's razor to clear away the theoretical deadwood, and they seem uncomfortable with theoretical or explanatory pluralism. However, evolution is not a simple phenomenon, and it seems preferable to accept that speciation may be allopatric or sympatric or some combination of both, that it may be both genic and chromosomal, that it may sometimes be gradual and sometimes saltational and sometimes a little bit of both, and that both anagenesis and cladogenesis and microevolution and macroevolution and all these other processes may be going on simultaneously. Rather than taking an adamant stand in favor of one or another competing theory, we might benefit from going back to studying the nature of species and speciation with open eyes and an open mind in order to determine the relative importance in nature of the many possible modes of speciation.

Chapter 12
Evolutionary Paradoxes

In recent writings about evolution, the word "paradox" has been prominent, in the sense of a statement that seems contradictory, unbelievable, or absurd, or opposed to common sense, but may actually be true in fact. Among these paradoxes are the paradox of sex, the paradox of the sex ratio, the paradox of meiosis, the paradox of recombination, the paradox of variability, and the paradox of adaptation. Thus some fundamental biological phenomena are thought to be paradoxical.

These paradoxes stem from the costs supposedly associated with each of the phenomena, which, according to prevailing evolutionary theory, should lead to the suppression or elimination of the phenomena by natural selection. Thus we read of the cost of sex, the cost of males, the cost of meiosis, the cost of recombination, the cost of fertilization, the cost of variability (otherwise known as the genetic load), and even the cost of natural selection and of evolution itself. Natural populations seem to be unfamiliar with the theories, however, for they ordinarily consist of equal numbers of males and females blithely engaging in sexual reproduction instead of consisting exclusively of asexually reproducing parthenogenetic females, as theory is thought to dictate.

The literature in this area is voluminous, but no effort will be made here to review it exhaustively. The following references will provide a sampling of the concepts and an entrée into the relevant literature (Fisher, 1930, 1958; Hamilton, 1964, 1967; Williams, 1966, 1971, 1975; Campbell, 1972; Ghiselin, 1974; Maynard Smith, 1978; Bell, 1982; Charnov, 1982; Trivers, 1985; Michod and Levin, 1988).

The costs and their attendant paradoxes result from the assumptions underlying the theories. In some cases the assumptions are explicit, in others they are implicit, but in either case the paradoxes are inherent in the assumptions rather than in the phenomena themselves. If different, more valid assumptions are made, the paradoxes vanish. An examination of the theories and especially of the assumptions on which they are based will show that the supposed paradoxes are false.

The following assumptions are so widely believed that they often remain unstated or else are given as universally accepted axioms:

1. Asexual reproduction is the primitive, ancestral mode of reproduction from which sexual reproduction has been derived.
2. Asexual reproduction is more efficient than sexual reproduction because there is a twofold cost to sex: the cost of males and the cost of meiosis.
3. Genetic recombination lowers fitness.
4. Inbreeding causes inbreeding depression.
5. Parental investment in male and female progeny must be equal; that is, the cost of producing males must equal the cost of producing females.
6. The reproductive value of offspring is equivalent to the parental investment in them.
7. Natural selection and evolution have a cost.
8. Finally, the simplest and most widely used assumption of all is *ceteris paribus*, that is, all other things being equal.

A welter of theories and models has been proposed to explain the seeming paradoxes. They include one of the most colorful assemblages of names known to science and seem to follow Bell's (1982) dictum, "We need names for theories, and preferably striking ones, if we are to avoid an enervating circumlocution every time we want to refer to an idea." A sampler would include the Tangled Bank; the Haystack model; the Hitchhiker effect; Muller's rachet; the Best Man hypothesis; the Runt model; the Red Queen effect; the Lottery Ticket model; the Elm-Oyster, Aphid-Rotifer, and Strawberry-Coral models; and the Vicar of Bray. These names are certainly striking and memorable, but in most cases, not particularly instructive.

Asexuality and Sexuality

Rather than dealing with this tangled bank of theories, let us discuss the underlying assumptions used so often in evolutionary ecology. The first assumption, that asexuality is ancestral and sexuality was derived from it (e.g., Maynard Smith, 1971, 1978), stems from the notion that the most primitive organisms, the prokaryotes, reproduced asexually. Sexual reproduction is supposed to have evolved later, imposing regularities on the distribution of chromosomes and genes to the progeny. However, the discovery of a variety of modes of genetic recombination in prokaryotes (e.g., transformation, transduction, and conjugation in bacteria) requires that this assumption be reexamined.

Before proceeding further, we must clarify the meaning of some of the terms being used, for common words such as sex often go undefined and may mean different things to different people. The essence of sex is genetic recombination; if genetic recombination occurs, a sexual process is involved.

Moreover, even though the words "sexual" and "reproduction" may seem inseparable, sex and reproduction are separate and distinct phenomena. Sex (i.e., genetic recombination) can occur without reproduction, and reproduction (i.e., an increase in the number of cells or individuals) can occur without sex.

Sex without reproduction occurs, for example, in the Protozoa by processes known as automixis and amphimixis. Mixis literally means mingling, but more precisely refers to reorganization of the genome giving rise to new combinations of genes. Conjugation in *Paramecium* is an example of amphimixis without reproduction. Conjugating paramecia, each with two micronuclei generated by meiosis, exchange a micronucleus with each other so that the ex-conjugants become genetically different from their original genotypes. Automixis is a modification of amphimixis that occurs within a single cell. The identical haploid micronuclei, instead of being exchanged with another cell, fuse with one another to produce a homozygous individual genetically distinct from the original cell. In both amphimixis and automixis new combinations of genes are generated with no increase in cell number (Sleigh, 1973); thus sex has occurred without reproduction.

It pays to be finicky about definitions because sloppiness can lead to confusion and verbal quagmires. Two examples will suffice. A prime case in point is the word "bisexual," which has two diametrically opposed meanings in biology (Rieger, Michaelis, and Green, 1976), the first being the equivalent of hermaphroditic or monoecious, and the second referring to populations composed of individuals of separate sexes, male and female. A third recent behavioral usage need not concern us here. More to the point is Bell's (1982) definition of genetic recombination, which he equates (pp. 20 and 506) with genetic crossing over between loci on the same chromosome. This idiosyncratic definition completely ignores Mendelian recombination resulting from the segregation and independent assortment of whole chromosomes, the other major cause of genetic recombination in higher organisms. Such an egregious error about a matter so fundamental to a book on the evolution and genetics of sexuality puts the entire work on a rather shaky foundation.

Meromixis refers to genetic recombination in which only a part of the genome of a donor cell is transferred to a recipient cell to form a partial diploid. Transformation, transduction, and conjugation in bacteria usually result in meromixis, and thus fall under the rubric of sex. In fact, sex is best defined as any process leading to the generation of new combinations of genetic material. Even in higher organisms genetic recombination by means other than Mendelian recombination and crossing over is being reported. The origin of recombination has recently been tied to DNA repair so that sex seems best defined broadly in terms of all forms of genetic recombination. If so, then asexuality as the primitive state from which sexuality has been derived is no longer a tenable hypothesis, for sexuality and asexuality seem equally ancient.

Reproduction without sex occurs not only in vegetative growth, budding, and fission but also in those forms of parthenogenesis that do not involve genetic recombination. Sexual reproduction is often equated with the union of gametes from individuals of opposite sex. However, separation of the sexes is not a necessary concomitant to sexual reproduction, for sexual reproduction (that is, genetic recombination accompanying reproduction) also occurs in hermaphrodites and monoecious plants. Therefore, "sex" must be distinguished from "gender," which refers to separate sexes, male and female. Even though species with separate sexes ordinarily engage in sexual reproduction, separate sexes are not essential for sexual reproduction. Conversely, asexual reproduction involves, not the absence of separate sexes, but the absence of genetic recombination. The distinction between sexuality and asexuality is the presence or absence of genetic recombination.

In 1971 Maynard Smith wrote, "In this essay I shall ask what selective forces were responsible for the origin of the sexual process, and by what selective process is it maintained." Implicit in this statement is the assumption that asexuality was the ancestral condition from which sexuality was derived, an assumption that has become embedded in textbooks (Stansfield, 1977, p. 145; Patterson, 1978, p. 167; Solbrig and Solbrig, 1979, p. 16; Futuyma, 1979, p. 307; Stanley, 1979, p. 224). The discovery of genetic recombination in viruses and bacteria suggests, however, that sex has existed since the origin of life and that sexuality and asexuality are of equally ancient vintage, a view expressed by Haldane (1954), Dougherty (1955), Stebbins (1960), Mayr (1963, p. 412), Ghiselin (1974, p. 64), and Williams (1975, p. 111).

Thus, it appears that the first assumption—that is, asexuality is primitive—is wrong. We no longer need ask why sexuality evolved from asexuality, but instead must ask what has led to the retreat from sexuality in so many groups of organisms and why sexuality has persisted if in fact it inflicts a double whammy, the cost of males and the cost of meiosis.

The Cost of Males

The second axiom, that asexuality is more efficient than sexuality, is based on this supposed twofold cost of sex. Let us first consider the cost of males. The argument is that if, in a sexual species with equal numbers of males and females, a mutation causing females to reproduce parthenogenetically occurs, it will sweep through the population rapidly, with the proportion of parthenogenetic females virtually doubling each generation at the outset (Maynard Smith, 1971, 1978). The reasoning is that if all available resources are devoted to the production of females and none is wasted on males, the parthenogenetic females will win out in competition with the sexual females because they can produce twice as many daughters as the sexual females, whose progeny are half daughters and half sons. Thus, natural selection will favor asexuality over sexuality. This reasoning has been applied not only to dioecious species but to hermaphrodites in which the resources are invested exclusively in eggs, with none wasted on sperm.

However, the idea that a mutation favoring female parthenogenesis will inevitably sweep through a population is simply not true. The fly in the ointment is the ubiquitous assumption, *ceteris paribus,* all other things being equal. The sad truth, as pointed out earlier, is that all other things are never equal (Lewontin, 1979a).

In the case of sexual versus parthenogenetic females, the things not equal are the genotypes of the progeny. The progeny of sexually reproducing females are genetically variable; the progeny of parthenogenetic females are genetically uniform. Maynard Smith (1978) assumed that the sexual and asexual females produced equal numbers of eggs with comparable survival rates, but such an assumption is contrary to all that is known about natural selection. The critical criterion of fitness is not fecundity or the number of daughters produced, but the proportion of the progeny surviving to reproduce. If the genetically variable progeny of sexual females survive and reproduce better than the genetically uniform progeny of the asexual females, the mutation producing parthenogenetic females will not sweep through the population but will be lost. Thus the argument for the cost of males assumes that fecundity is equivalent to fitness and that all other things are equal. Since neither assumption is true, the argument fails. There is no inherent cost of males in sexual reproduction.

Surprisingly, Lewontin (1979b) wrote in a similar vein, "A mutation which doubles the fecundity of individuals will sweep through a population rapidly," and a few sentences later, "natural selection at all times will favour individuals with higher fecundity." Again *ceteris paribus* was assumed. However, if fecundity is doubled, the resources used for egg production need to be doubled, or else the size of the eggs must be halved, or a doubling in the efficiency of resource utilization must occur, but in any case all else would not be equal. The work on clutch size in birds (Lack, 1954, 1966) has shown that selection does not favor unlimited fecundity but rather favors a clutch size that permits a maximum number of young to be fledged. Thus selection may act against high as well as low fecundity, and a gene for high fecundity will not necessarily sweep through a population, for the ultimate test of this gene is its effect on reproductive fitness, not fecundity.

The Cost of Meiosis

The other half of the supposed twofold cost of sex is the cost of meiosis. In this case the reasoning is that an asexual female transmits all of her genes to each of her offspring whereas a sexual female transmits only half of her genes to each offspring. The diploid eggs of a female reproducing parthenogenetically contain her complete genome; the haploid eggs of a sexual female contain only half of her genes following meiosis, with the other half of the genes in the diploid progeny coming from the male at fertilization. Thus, the genes of a parthenogenetic female should be twice as well represented in the next generation as those of a sexual female, all other things being equal. As Williams (1971) wrote, "Meiosis is therefore a way in which an individual actively reduces its genetic representation in its own offspring."

Again, all else is not equal because the progeny of asexual and sexual females have different genotypes. If the sexually reproducing females make a greater genetic contribution to the breeding population of the next generation than the asexually reproducing females, they will have greater fitness. The test of fitness in this case is not what proportion of her genome a female transmits to each of her offspring, but how well her genes are represented in the gene pool of the next generation. Thus, the cost of meiosis, like the cost of males, is based on a false premise, *ceteris paribus,* and on the assumption that fecundity is equivalent to fitness. As neither is true, the argument for the cost of meiosis also fails.

Depending on the conditions, either asexuality or sexuality or even some combination of both may be favored by natural selection. In any case there is no intrinsic twofold advantage of asexual over sexual reproduction. Fitness is more than a numbers game, measured in terms of fecundity or reproductive potential; greater fecundity does not guarantee greater fitness. To assess the relative fitness of sexuality versus asexuality, the relative genetic contributions to the gene pool of the next generation by sexual and asexual individuals must be measured.

The adaptive significance of asexuality and sexuality may be inferred from studies of such animal species as cladocera, rotifers, and aphids and the many species of plants in which facultative sexual and asexual reproduction occurs. In general, asexual reproduction predominates when environmental conditions are favorable and stable, and well-adapted genotypes can proliferate rapidly with little wastage of poorly adapted progeny. Sexual reproduction ordinarily intervenes when conditions become unstable or unfavorable, thus increasing the chances that among the array of genotypes produced, some will be able to adapt to the changing environmental conditions.

Finally, both Williams (1971, 1975) and Maynard Smith (1971, 1978) thought sexuality posed a paradox because although it facilitated long-range evolutionary adaptation by means of group selection, it was detrimental to the reproductive interests of individuals because of its twofold cost. However, there is no need to postulate long-range evolutionary advantage or group selection to account for the evolution and maintenance of sexuality. Individual selection and the immediate adaptive advantages of sexuality seem sufficient to explain both its persistence and its prevalence.

The Cost of Recombination

Next let us consider the assumption that genetic recombination lowers fitness. The argument seems to be that the combinations of genes in living organisms are survivors of the evolutionary process, the most favorable having been preserved by natural selection, and that genetic recombination will break up these favorable gene complexes, leading to a loss of fitness. This loss of fitness is the cost of recombination, and as such poses the paradox of recombination. With such a cost, strong selection against genetic recombination would be expected, and, as Turner (1967) put it, the genotype should congeal. Although there is considerable evidence that genetic recombination is influenced by selection, recombination is far from being eliminated by natural selection. Despite the supposed cost of recombination, gene complexes have not been locked up in congealed genotypes, and genetic recombination persists.

The cost of recombination seems to be based on at least two assumptions that need scrutiny. The first is that the existing well-adapted genotypes, survivors of selection, will be equally well adapted in subsequent generations, in other words, that the physical and biological environment is constant. This simplifying assumption, though appealing, is quite unrealistic; as pointed out earlier, the environment is in a continual state of flux, and the better-adapted genotypes will change as the population tracks the shifts in environmental conditions. Recombination, by permitting the continuous generation of new genotypes, is an adaptive device that keeps the individuals in the population from being locked into outmoded genotypes and thus should be favored by natural selection.

Furthermore, favorable gene complexes are not inevitably broken up each generation, for recombination is neither random nor unlimited. The amount of genetic recombination is regulated by natural selection through its effects on the frequency of chiasma formation, the size of chromosomes, the frequency of chromosomal rearrangements, and in other ways. The rate of release of genetic variability in natural populations can thus be fine-tuned by natural selection. Rather than postulating a cost of recombination, one can just as well argue that a lack of recombination carries a cost in a changing environment.

The other assumption underlying the cost of recombination is that the genotypes of the organisms are constant, that these well-adapted genotypes are fixed and unchanging. However, the estimated whole chromosome mutation rates for major deleterious mutations are of the order of 1 percent per chromosome per generation (Dobzhansky, Spassky, and Spassky, 1952, 1954). If mutations of lesser effect are included, it is clear that the fitness of genotypes is constantly subject to erosion from the effects of mutation. With asexuality, the accumulation of deleterious mutants goes on unchecked, a process dubbed Muller's rachet. In an asexual line, once a harmful mutation occurs, the only way the line can get rid of it is by means of an improbably rare back mutation. Apart from this an asexual population will never come to contain, in any of its lines, a line with a load of mutations lower than the least-loaded existing line. Mutation will only continue to turn up the rachet additional notches.

Therefore, sexual reproduction has a clear advantage over asexual reproduction because genetic recombination provides a means of eliminating harmful genes from the population and of negating the effects of Muller's rachet. The other side of this coin is that genetic recombination also permits the formation of new gene combinations, some of which may be more favorable than any existing combinations. Crow (1988), in discussing the evolutionary importance of recombination, cited as the three major advantages of sexual reproduction: (1) adjusting to a changing environment, (2) incorporating beneficial mutations in a single lineage, and (3) getting rid of deleterious mutations. He seems to lean toward the third, the reduction of genetic load, as the most important of the three. This opinion is valid if based on the assumption of constant fitnesses in constant environments. However, beneficial genes may originate not just by mutation *de novo*, but as the result of new epistatic interactions or from shifts in adaptive value following environmental change. Cleansing the gene pool of deleterious genes may be an important role for sexuality, but the ability to form new adaptive gene combinations and to respond to changing environmental conditions seem even more important. The predominance of sexuality in the more complex, more highly evolved forms of life suggests a more fundamental role for sex than merely eliminating harmful genes.

Inbreeding Depression

In the literature on the evolution of sex, the word "inbreeding" is often followed by the expression "inbreeding depression," as though one inevitably follows the other (e.g., Charnov, 1982; Trivers, 1985). This assumption, of course, is not true. Inbreeding does not cause inbreeding depression; inbreeding causes an increase in homozygosity. If the increased homozygosity leads to an increase in the proportion of deleterious recessive homozygotes in a population compared to previous generations, there is a concomitant reduction in overall fitness and inbreeding depression can be said to have occurred. However, high levels of inbreeding, including selfing, are the norm in many species, giving rise to vigorous, fertile offspring, so there is nothing inherently harmful about inbreeding. Inbreeding depression is ordinarily observed when previously outbreeding populations are inbred. This seemingly innocuous assumption about the association between inbreeding and inbreeding depression can cause considerable mischief if used as the basis for speculations about the evolution of sex and of mating systems.

The Sex Ratio, Parental Investment, and Reproductive Value

First, a few words about the sex ratio may be in order. The sex ratio is a population characteristic, but, like many other group characters, it is influenced by individual selection rather than group selection. The sex ratio in a cohort may be estimated at fertilization, at birth or at hatching, at weaning, at the time parental care ceases, at sexual maturity, or at any other time for that matter. The ratio of males to females at fertilization is known as the primary sex ratio; sex ratios at later stages may be called secondary. If one attempts to determine the sex ratio in a population using individuals of all ages, the potential difficulties should be obvious.

It should be added that sex may be determined in a variety of ways, ranging from chromosomal to environmental. Sex determination usually occurs early in development, but if males and females subsequently die at different rates, the sex ratio will not remain constant but will vary. Furthermore, genetic mechanisms are known that can modify the sex ratio (Edwards, 1962; Eshel, 1975; Maynard Smith, 1978), so that the sex ratio, like any other trait, is subject to the effects of natural selection. A primary sex ratio, for example, chromosomally determined at fertilization, can be modified or fine-tuned subsequently by natural selection.

Earlier it was asserted that although the sex ratio is a group characteristic, it is controlled by individual selection. The reasoning is relatively straightforward. The zygotes giving rise to the next generation come from a sperm fertilizing an egg; thus, half the gametes come from the male parents and half from the females. For example, 100 offspring come from 200 gametes, 100 sperm, and 100 eggs. If the parents consisted of 10 males and 20 females, then each male contributed 100/10, or 10 gametes per male on average, and each female only 100/20, or 5 gametes per female, to the next generation. Therefore, each male averaged twice as great a genetic contribution to the next generation as each female, and any individual in the population carrying genes that shift the sex ratio in its progeny toward an increased proportion of males would be favored by natural selection. Conversely, if males are in excess, natural selection would favor individuals carrying genes that shift the sex ratio in their progeny toward a greater proportion of females. In either case, as the sex ratio approaches unity, such genes lose their selective advantage, for the sex ratio is a balanced frequency-dependent trait. Because genes tending to produce the rarer sex will always be favored by selection, natural selection will tend to drive the sex ratio toward a 1:1 ratio at the age of reproduction, as Fisher first pointed out in 1930.

Next we must consider parental expenditure or parental investment, otherwise known as the biological capital invested by parents in their offspring or the cost of producing males and females. This concept stems from R. A. Fisher (1930, 1958) in a brief section headed "Natural Selection and the Sex-Ratio." Although authors generally credit Fisher for initiating research in this area, their praise is usually qualified (e.g., Crow and Kimura, 1970): "This problem (the 1:1 sex ratio) was first solved by R. A. Fisher (1930). His treatment involved the concept of 'parental expenditure,' the meaning of which is not entirely transparent (to us, at least)." Charnov (1975) stated that the notion of parental resources and investment is "one of the most obscure concepts in sex ratio theory," and in the same vein, Bull and Charnov (1988) wrote, "It is an understatement to suggest that Fisher's argument is cryptic; various models have since been published that merely attempted to elucidate Fisher's argument, attesting to the difficulty of the problem."

It is worth noting that despite the twenty-eight years between the 1930 and the 1958 editions of *The Genetical Theory of Natural Selection*, Fisher did not see fit to change a single word in the section entitled "Natural Selection and the Sex-Ratio." Given the confusion caused by this section, it would be interesting to know why he never revised it.

Fisher expressed his argument in economic terms such as biological capital, parental expenditure, and reproductive value, leading one to expect a quantitative treatment of parental investment. Economic change, however, can be measured in terms of dollars and cents; biological expenditures are not so easily measured. Fisher indicated that parents invest in their offspring a certain amount of nutriment plus an added amount of time or activity but did not reveal how to find a common unit by which parental investments in nutriment, time, and activity could be measured.

Fisher's successors have followed his lead on parental expenditures, but expenditures are easier to define than to measure, and they are not particularly easy to define. Some of the proffered definitions are as follows. Trivers (1972) called parental investment "any investment by the parent in an individual offspring that increases the off-spring's chance of surviving (and hence reproductive success) at the cost of the parent's ability to invest in other offspring." In parental investment he later (1985) included "metabolic investment in the primary sex cells" as well as "feeding or guarding the young" and any other time or effort expended on behalf of the progeny. Trivers also at various points referred to parental investment in terms of resources, work, energy, risk, and time, but the means of making objective measurements of parental investments in such disparate traits as risk and energy were not given. Bodmer and Edwards (1960) referred to "parental expenditure of effort"; Emlen (1968a, b) wrote of parental expenditure in terms of "energy" expended; and Hirschfield and Tinkle (1975) defined reproductive effort as the proportion of the total energy budget devoted to reproductive processes. Wilson (1975) wrote that "effort expended on reproduction is not to be measured directly in time or calories. What matters is benefit and cost in future fitness." Apparently because the measurement of future fitness presents obvious problems, Wilson also wrote: "It makes sense to describe reproductive effort in terms of its physiological and behavioral enabling devices, such as the proportion of somatic tissue converted to gonads and the amount of time spent in courtship and parental care. However, the performance of these devices must be converted into units in the life tables before their effects on genetic evolution can be computed." Wilson presented no method for making such conversions, however.

Some efforts have been made to measure parental investment. Bull and Charnov (1988), for example, wrote that "investment has typically been calculated from dry weights of the reproductive individuals at the time of emergence." Just how risk or time is to be factored into such measurements is a puzzle. Perhaps Bull and Charnov's earlier statement best reflects the situation: "One of the difficulties in considering whether sex ratios satisfy equal investment is that the relative cost of a son versus a daughter has virtually never been measured properly. For the most part, only qualitative estimates of differential cost have been proposed." Therefore, it seems obvious that the concept of parental expenditure remains quite nebulous, at best, and that the costs of producing males and females have not been adequately measured. Nevertheless, despite the absence of reliable data, sex ratio theory has been built on the idea that the equality of the sex ratio is dependent on an equal allocation of resources between males and females.

The sixth assumption listed for scrutiny is that the reproductive value of offspring is equivalent to the parental investment in them. This assumption is perhaps the most mischievous of all. Let us review what Fisher (1958) wrote: "In organisms of all kinds the young are launched upon their careers endowed with a certain amount of biological capital derived from their parents. . . . Let us consider the reproductive value of these offspring at the moment when this parental expenditure on their behalf has just ceased. If we consider the aggregate of an entire generation of such offspring it is clear that the total reproductive value of the males in this group is exactly equal to the total value of all the females, because each sex must supply half the ancestry of all future generations of the species. From this it follows that the sex ratio will so adjust itself, under the influence of Natural Selection, that the total parental expenditure incurred in respect of children of each sex, shall be equal; for if this were not so and the total expenditure incurred in producing males, for instance, were less than the total expenditure incurred in producing females, then since the total reproductive value of the males is equal to that of the females, it would follow that those parents, the innate tendencies of which caused them to produce males in excess, would, for the same expenditure, produce a greater amount of reproductive value; and in consequence would be the progenitors of a larger fraction of future generations than would parents having a congenital bias towards the production of females. Selection would thus raise the sex-ratio until the expenditure upon males became equal to that upon females."

The crucial phrase in this passage is "From this it follows," where Fisher states that because the total reproductive values of males and females are equal, the parental expenditures on males and females must also be equal. Although the logic of the equality of total male and female reproductive values is inescapable, it by no means necessarily follows that parental expenditures on males and females must also be equal. This type of reasoning brings to mind the confounding of fecundity with fitness. High fecundity is not necessarily related to high fitness nor do equal parental expenditures on each sex necessarily lead to equal reproductive values for males and females. The important factor is that the reproductive values of males and females are equal, not that parental expenditures on males and females must be equal.

In many species, males and females differ in size. For the sake of argument, let us suppose that, in a certain species, a female is twice as large and twice as costly to produce as a male. If natural selection is working to produce equal parental expenditures on males and females, then the expected sex ratio in this species is two males to one female. However, if natural selection is working to produce equal reproductive values in males and females, the sex ratio should be one male to one female despite their differences in size and in cost to their parents. The fundamental question is whether natural selection acts on parental expenditures in order to equalize the reproductive value of males and females or whether it can equalize the reproductive values by acting directly on the sex ratio. An examination of the sex ratios in species with markedly different males and females and presumably different costs per male and female should help to resolve this question.

Ever since Darwin (1871) it has been generally agreed that sex ratios in natural populations of dioecious species are typically close to 1:1 (Hamilton, 1967; Crow and Kimura, 1970; Hartl and Brown, 1970; Eshel, 1975; Trivers, 1985) even though reliable data on sex ratios in natural populations are, for a variety of reasons, difficult to obtain (Trivers, 1972). If sex ratios are determined by equal parental expenditures on males and females, the generality of the 1:1 sex ratio is quite surprising. If selection tends to equalize parental expenditures on males and females, the total investments in males and in females will be nearly equal, but the sex ratio will be skewed whenever expenditures per male and per female are unequal. Therefore, if equal total expenditures on males and on females are the ultimate determinant of the sex ratio, a much wider range of sex ratios would be expected than is actually observed.

On the other hand, if sex ratios are dependent on the equal reproductive values of males and females, then frequency-dependent selection will always tend to drive the sex ratio toward 1:1, no matter what the parental expenditures on males and females may be. The fact that sex ratios in natural populations so often approximate a 1:1 ratio suggests that the ultimate determinant of sex ratios in natural populations is not parental expenditures but the equal reproductive value of males and females. Because the distribution of sex ratios expected under the two concepts is quite different, it should be possible to decide which is true. However, the lack of reliable data on parental expenditures and on sex ratios in natural populations still leaves room for argument. Nevertheless, the preponderance of 1:1 sex ratios in wild populations suggests that a form of balancing frequency-dependent selection based on the equal reproductive value of males and females is responsible for the observed sex ratios.

This line of reasoning leaves the matter of parental investment in limbo. Parental investment varies markedly in different species, in some cases being essentially zero, apart from the expenditures on the gametes. In other species the investments in the young are more diverse, including not only the gametes but also feeding, guarding, instructing, and otherwise helping the young to reach maturity. There seems to be an inverse relation between numbers of progeny and the amount of parental expenditure per offspring; the greater the investment per offspring, the fewer the number of progeny produced. A pair of eagles produces one to three eggs per year and devotes considerable care to their young, whereas oysters discharge millions of sperm and eggs into the sea but provide them with no care at all. This relation indicates that there must be limits on the amount of capital that parents can afford to invest in their progeny and that selection must control the way in which these resources are invested, but it does not necessarily indicate that these resources need be equally allocated between males and females. Selection, as stated earlier, will act to maximize fitness, that is, to maximize the genetic contribution of individuals to the gene pool of the next generation. This maximization is achieved when a form of balancing frequency-dependent selection produces equal numbers of males and females at reproductive age. The overriding selective consideration is the equal total reproductive value of the males and females at that time; parental expenditures on the two sexes are secondary. Thus, parental expenditures will be adjusted, not to be equal for males and for females, but to produce equal numbers of males and females. If the costs per male and per female are different, then total parental expenditures on males and on females may well be different.

Furthermore, if the sex ratio were in fact determined by parental expenditures on males and females, then adjustments of the sex ratio to compensate for differential mortality in the sexes during development would have to cease when parental care ceased. However, the 1:1 sex ratio needed to provide equal reproductive values is essential not at fertilization or at birth or at the end of parental care, but at the time of reproduction. In many species, considerable time may elapse between the end of parental care and the start of reproduction. Any differential mortality that occurred during this period could no longer be compensated by parental expenditures.

Fisher recognized this problem and addressed it in the final paragraph of his section "Natural Selection and the Sex-Ratio." He wrote as follows: "The sex-ratio at the end of the period of expenditure thus depends upon differential mortality during that period, and if there are any such differences, upon the differential demands which the young of such species make during their period of dependency; it will not be influenced by differential mortality during a self-supporting period; the relative numbers of the sexes attaining maturity may thus be influenced without compensation, by differential mortality during the period intervening between the period of dependence and the attainment of maturity." Having written all this, Fisher then continued: "Any great differential mortality in this period [from the end of dependence to maturity] will, however, tend to be checked by Natural Selection, owing to the fact that the total reproductive value of either sex, being, during this period, equal to that of the other, whichever is the scarcer, will be the more valuable, and consequently a more intense selection will be exerted in favour of all modifications tending towards its preservation." In this rather lengthy and involved sentence, he further modified his position, for he now stated that even after parental care ends, natural selection will tend to counteract any great differences in mortality between the sexes because of their equal reproductive value at the time of reproduction. As selection on the sex ratio cannot operate through modifications of parental expenditures after parental care has ceased, it must work through other genetic means.

In certain mutant stocks of *Drosophila* maintained in the laboratory, it was observed that the sex ratio consistently showed an excess of males whereas in others the females were in excess (Merrell, unpub.). Although the exact mechanisms responsible for these differences are unknown, the implications are clear. Genes exist in *Drosophila* that can modify the chromosomally determined sex ratio. The primary 1XX: 1XY sex ratio at fertilization is based on the random segregation and combination of chromosomes during meiosis and fertilization. This primary sex ratio could subsequently be modified by selection on modifying genes to ensure that a 1:1 sex ratio and equal reproductive values for males and females will exist at the time of reproduction.

The preceding arguments, based on the equal reproductive values of the sexes, apply to dioecious diploid species without regard to the actual mode of sex determination. In other words, they apply to such species even when sex is determined environmentally. The triggering environmental stimulus may be temperature as in turtles or the presence or absence of a female as in the marine worm *Bonellia*, but the ability to respond to such stimuli is genetically controlled and thus can be subjected to selection like other types of traits.

Social Evolution

Much work on the sex ratio and on kin selection has been done on haplodiploid social insects in the Hymenoptera where sex determination depends on whether the female lays an unfertilized egg (male) or a fertilized egg (female). In the theoretical discussions it has been assumed that equal parental expenditures rather than equal reproductive values determine the sex ratio. However, Hartl and Brown (1970) reported that "haplodiploid populations tend towards and maintain equilibrium sex ratios of one-to-one, despite the grossly unequal genetic contributions made by the two sexes to the developing eggs of each new generation," which, they averred, came to them as "a mild surprise." Thus, even in haplodiploids, where the sex ratio might be expected to deviate widely from 1:1 because of (1) the method of sex determination, (2) the differences in parental expenditures, and (3) the minor role played by the males in the life of the colony, the sex ratios approximate 1:1. Such ratios in haplodiploids are more readily understood in terms of the equal reproductive values of males and females than in terms of equal parental investments in males and females.

It has been further assumed that the queens mate only once, which led to the conclusion that sisters share three-fourths of their genes in common and are more closely related to each other than they are to their brothers or their parents. From this supposition further conclusions have been drawn about kin selection. The difficulty with this assumption is that queens in honeybees, for example, are known to mate with about ten to twenty drones rather than just one (Adams et al., 1977; Laidlaw and Page, 1984; Collins et al., 1987). Thus, the degrees of relationship among the queens, drones, and workers are not as postulated. Laidlaw and Page reported, for example, that the average genetic relatedness among honeybee nestmates is low and approaches 0.25 in colonies with naturally mated queens. Thus, the theoretical foundation underlying studies of the sex ratio and kin selection in haplodiploids involves assumptions that need review.

Social evolution in the Hymenoptera has also been characterized by discussions of parent-offspring conflict, sibling rivalry, and conflict between workers and queens (Trivers, 1985). In the latter case, the workers, which are nonreproductive females, are treated as independent players in the selective game. However, they are more properly regarded as part of the extended phenotype of the sexual founders of the colony. The success or failure of colonies, to the extent that it is genetic, is dependent on the genetic qualities of the queens and drones generating the colonies. Some queens and drones will generate more successful colonies than others, in part because they produce better workers. The workers, however, are not subjected to natural selection directly because they do not normally reproduce. Therefore, they can hardly successfully push their own independent selective agenda, for their contributions to the success of the colony can only be passed on through the reproductive members of the colony.

In the case of parent-offspring conflict, both parents and offspring have a common interest in the survival of the offspring. The reasons for parent-offspring conflict in mammals at weaning and birds at fledging may lie not in some subtle weighing of the costs and benefits of different reproductive strategies, but in some simpler selective explanation such as the need for parents to return to devoting all of their dietary intake to their own maintenance in preparation for migration or survival during the winter. This is not to say that all explanations of behavior invoking genetic cost-benefit analysis or some type of kin selection should be dismissed out of hand (the genetic explanation for the murder of lion cubs by new males taking over a pride seems well validated [Schaller, 1972]), but rather that such thinking should not be universally applied when other viable explanations exist.

The Cost of Natural Selection

Now we must turn to the assumption that there is a cost to natural selection and evolution. The concept of the cost of natural selection was first handled quantitatively by J. B. S. Haldane (1957, 1960). What emerged from his analysis posed another paradox, known as Haldane's dilemma, because the cost of natural selection appeared to be so great that, for the cost to stay within reasonable limits, the rate of evolution had to be extremely slow.

Haldane framed the problem in simple terms: How many selective deaths must occur in a population while one allele is replacing another? The newly favored rare allele may arise either by mutation or because of a change in environment. Haldane calculated that a simple gene substitution was extremely costly even under the most favorable conditions, requiring a cost in genetic deaths of at least ten to twenty times the number of survivors in a single generation and probably more. To make an exact estimate, he assumed a cost of 30 and also that 10 percent of the reproductive excess of each generation could be devoted to this allelic substitution. With these assumptions, he estimated that a species could carry out only one allelic substitution every 300 generations on average. Because even closely related species differ at many loci, these findings suggested that, because of their cost, substitutions leading to new species must have occurred sequentially and had to be spread over many thousands of generations.

This supposed burden on the population, dubbed the substitutional load by Kimura (1960), was a major reason for Kimura's (1968) proposal of the neutral mutation-random drift hypothesis of molecular evolution. He calculated the substitutional load to be so great that no mammalian species could tolerate it. However, if the alleles were assumed to be selectively neutral, this costly substitutional load automatically disappeared. Further discussion about estimates of the cost of natural selection can be found in Crow (1970), Crow and Kimura (1970), Merrell (1981), and Kimura (1983).

Despite the clarity of the calculations and conclusions about the cost of selection, something is obviously amiss, for evolution by selection can sometimes be very rapid. The evolution of industrial melanism in many species of moths and of insecticide resistance in many insect species are usually mediated by single-gene substitutions and have occurred within a few rather than hundreds of generations. Moreover, all the many allelic substitutions that distinguish domesticated animals and plants from their wild ancestors have occurred in some 10,000 years or less, again suggesting that selection cannot be nearly as costly as the calculations indicate. If not, then the methods of estimating costs must be examined.

Briefly, the calculations showed that the number of generations required for a given change in allelic frequency was inversely proportional to the intensity of selection. The less intense the selection, the greater the number of generations required to effect a given change. Moreover, the total cost is less if the selection coefficient (s) is small. However, the cost does not change much when s becomes as small as 0.1, and becomes essentially constant when s is as small as 0.01. At that point, the cost becomes independent of s but depends on the initial frequency of the favored allele. Tables of cost, constructed for various initial frequencies of the favored allele and for various values of s, showed enormous costs for natural selection and such a loss of fitness that the populations might be expected to go extinct rather than evolve.

However, these estimates of cost were based on the highly unlikely assumption that population size is roughly constant during the period of allelic replacement. In reality, population sizes in natural populations tend to show wide fluctuations. Although the possibility was recognized that a newly favored allele might arise because of an environmental shift as well as by mutation, this possibility was not really explored, and constant environmental conditions were assumed.

It was also assumed that selection coefficients were small and that genes at different loci acted independently.

The concept of the cost of natural selection or the substitutional load is rather slippery. One reason is that it deals with relative fitnesses. With this mode of thought, the appearance of a new favorable mutation in a population creates an immediate genetic load. The quickest way to get rid of the load would be to eliminate the new mutation. With this new mutation, however, the population is actually better off than it was before. Rather than being dragged down by a newly created load, it has, in absolute terms, a higher level of fitness than before.

The so-called cost of selection measured in terms of deaths is, in a sense, irrelevant. What is important is not how many individuals die, but how many and what kinds of individuals survive. The artificial assumptions of constant population size, a more or less constant environment, small selection coefficients, and independently acting loci facilitate the calculation of the cost of evolution, but are so far from the conditions confronting wild populations as to make the results virtually meaningless. Wallace (1991, p. 76) also feels that Haldane's calculations for the cost of evolution were flawed.

Populations fluctuate in numbers from season to season and from year to year. Selection pressures change from season to season and from year to year. The average population size and the average selection pressure are meaningless to a population confronted by a drought or an insecticide or the destruction of its habitat. The lucky or hardy or well-adapted few that survive these challenges become the progenitors of the subsequent generations of this population. If none survives, the population's fate is extinction, but if any survive, the number of deaths is not important, but the number of survivors and their characteristics are. The cost of not evolving, as Brues (1964) put it, is by any measure far greater than the supposed cost of evolution.

The cost of mutation, the cost of variation, the cost of recombination, and the cost of natural selection and evolution are all ideas that stem from the concept of genetic load. However, evolution in sexually reproducing species occurs because of mutation, variation, recombination, and natural selection and not in spite of them. There is a sort of inverted, twisted logic to the whole concept of the cost of evolution.

Numquam Ceteris Paribus

Finally, we must address again the most widely used assumption of all in evolutionary biology, *ceteris paribus*, all other things being equal. This assumption is usually made so that the effects of changing one factor in a complex situation can be evaluated while all others are held constant. Although this assumption has permitted many conceptual advances to be made, the great difficulty is that it is unrealistic, for in complex biological systems all else is never equal, that is, *numquam ceteris paribus*. As we saw earlier in the discussion on the supposed twofold cost of sex, changes in one factor inevitably led to changes in others. At present, *ceteris paribus* serves more as a crutch to theory than as a means to a better understanding of the mechanism of evolution. The difficulty is not so much that simplifying assumptions are made in the formulation of theory but that, once made, they seem to become a part of the theory, and the underlying assumptions on which the theory rests are ignored or forgotten.

Chapter 13
Behavior and Evolution

The word "behavior" is so common that it frequently goes undefined, even in books about behavior. Everyday notions of what constitutes behavior are often diffuse, so the nature of behavior requires some discussion. Behavior is difficult to define because it is not a structure but a sequence of events with both spatial and temporal dimensions. One definition is that behavior is what an organism does, involving action and response to stimulation, and may involve muscular and glandular activity. It has also been defined to include all those processes by which an animal senses the external world and the internal state of its body and responds to changes that it perceives. Another definition is that the study of behavior is the study of what animals do, and how and why they do it. At times, however, what they do not do can be as revealing as what they do. Behavior can be thought of as a strip map of the life of an organism, a joint function of the organism and its surroundings through time.

The Study of Behavior

The study of behavior has ancient roots, in zoology, in medicine, and in philosophy. Philosophy and medicine have always been primarily concerned with human rather than animal behavior. Of course, humans are animals, too, and their behavior, like that of other animal species, has an underlying biological basis. Scientists in many diverse disciplines study behavior; neurologists, psychiatrists, psychologists, behavior geneticists, physiologists, biochemists, endocrinologists, neurobiologists, ethologists, and sociobiologists are all students of behavior, but their approaches and their goals may be very different. Many, like the psychiatrists, concentrate on human behavior, particularly aberrant human behavior; others, such as the ethologists, for the most part ignore human behavior.

In general, three major approaches are used to study behavior: the way of the physiologist, the way of the psychologist, and the way of the naturalist. The way of the physiologist, employed by biochemists, cell biologists, endocrinologists, geneticists, and neurobiologists as well as physiologists is reductionist, an effort to delve into the genetic, neural, biochemical, hormonal, sensory, or muscular mechanisms underlying behavior. The way of most psychologists is to study intact animals, often under unnatural experimental conditions. Ethologists and sociobiologists take the way of the naturalist, a more holistic approach, seeking to study and understand the behavior of free, unfettered animals in their natural environments.

Two major types of behavior can be recognized: individual behavior and social behavior. Think, for a moment, of the kinds of individual behavior that have been identified (this list is by no means exhaustive): reflexes, fixed action patterns, sensory capacities, conditioning, habituation, learning, foraging, food selection, habitat selection, orientation, navigation, homing, migration, dispersal, geotaxis, phototaxis and other taxes, instincts, motivation, drive, emotional behavior, thermoregulation, play, and so on.

Social behavior can be defined as interactions between different individuals and thus may be either intra- or interspecific. Intraspecific social behavior includes courtship, mating, nesting, territoriality, parental care, imprinting, teaching and learning, communication, cooperation, altruism, colony formation, communal displays, flocking or schooling, play, migration, competition, aggression, dominance, threat, fighting, and displacement and appeasement activities.

Like some of the interactions between members of the same species, interactions between members of different species are often antagonistic and perhaps should be called antisocial rather than social. Species interactions include competition, predation, parasitism, commensalism, symbiosis, communication, and plant/animal interactions. Not all species interactions are easily categorized. A mixed species flock of migrating fall warblers that briefly enlivened a previously silent northern Minnesota woods, joined for a time in their feeding by the resident chickadees, comes to mind as an unexplained marvel of interspecific group behavior.

From this welter of examples of individual and social behavior, what inferences can be drawn? First, that behavior is a phenotypic character subject to the same evolutionary forces as the rest of the phenotype. Further, because behavior is a continuously unfolding trait, it is often necessary to identify certain aspects or units of behavior as suitable objects for study, for it seems impossible even to attempt to study the total lifetime behavior of an organism or group of organisms. Moreover, since behavior is a product of evolution, the evolution of behavior must have resulted from the same combination of adaptive and nonadaptive processes that has governed the evolution of other aspects of the phenotype. Our purpose, however, is not to study the evolution of the great diversity of types of individual and social behavior that have been mentioned.

Because behavior is a product of evolution, it can, properly used, be as helpful in building phylogenies as morphology or molecules. Again, however, our purpose is not to consider the construction of behavioral phylogenies even though they are often very informative. In the gray treefrog, for example, differences in the mating calls of the males and in the mating preferences of the females provided the first clues that morphologically similar populations of gray treefrogs actually consisted of a diploid species, *Hyla chrysoscelis*, and a tetraploid species, *Hyla versicolor* (Johnson, 1966; Ralin, Romano, and Kilpatrick, 1983). See also the role of behavioral studies in the systematic work on over 500 endemic species of Hawaiian Drosophilidae (Carson et al., 1970; Spieth, 1982).

Of the many types of behavior cited, most are, in one way or another, adaptive; that is, they contribute to the survival and reproduction of the organisms. Some types of aberrant human behavior may be self-destructive and maladaptive, but most behaviors, from foraging for food to reflexive withdrawal from fire, are functional and adaptive. Because, as noted earlier, the mechanism of adaptation is natural selection, the presumption is strong that adaptive behavior is produced by natural selection. To some, such a bald statement may seem to have a ring of heresy and probably requires amplification.

Nature and Nuture

The study of behavior has long been plagued and is still plagued by the nature-nurture controversy (witness the uneasy reception given to the findings of the Minnesota study of identical twins reared apart by Thomas J. Bouchard and his associates, 1990). This controversy has undoubtedly been most intense among students of behavior. On the one hand, there is the school of Behaviorism exemplified by J. B. Watson (1930), who in essence held that environmental factors were responsible for individual differences in behavior. The most famous statement of his position was: "Give me a dozen healthy infants, well-formed, and my own specified world to bring them up in and I'll guarantee to take any one at random and train him to become any type of specialist I might select—doctor, lawyer, artist, merchant-chief, and, yes, even beggar-man and thief, regardless of his talents, penchants, tendencies, abilities, vocations, and race of his ancestors." This environmentalist attitude toward behavior was prevalent for a long time among American psychologists, perhaps because so many of them were engaged in studying learning in the white rat (see Beach's delightful 1950 article "The Snark Was a Boojum"). In stark contrast were the ethologists, mainly in Europe, who studied fixed action patterns, innate releasing mechanisms, "instincts," and the like, and took the inheritance of behavior for granted.

This difference of opinion about the nature of behavior is far more than a scientific dispute, for it has political, religious, and social overtones as well. Some people are willing to accept a role for heredity in the behavior of other species, but seem to draw the line at the idea that heredity might have an influence on human intelligence, personality, or emotional stability. Given the simplistic ideas of the early eugenics movement and the horror caused by Hitler's racist perversion of genetics in his pogrom of the Jews, this attitude is understandable. However, humans are primates that, on the basis of molecular, cytological, and morphological evidence, are closely related to the great apes. To set humans apart as unique and distinct from all other animal species with respect to behavior ignores the evidence for this close relationship. Every species is unique in some way, but it has been argued that the uniqueness of humans is of a different order because in humans, biological evolution has been superseded by cultural evolution. Although human behavior and culture are undoubtedly extremely complex, they result from the activities of the human animal, a biological organism composed of flesh, blood, and DNA. The conclusion is inescapable that human behavior and culture must have a biological as well as an environmental basis. Nonetheless the nature-nurture dichotomy continues to plague the study of behavior, especially human behavior.

As discussed previously, the either-or, nature-nurture approach to behavior, as to other biological problems, is the wrong approach. The appropriate question is not whether behavior is determined by heredity or environment, but to what extent heredity and environment contribute to the observed behavior. Even that most dedicated of environmentalists, J. B. Watson, requested that he be provided with a dozen healthy, well-formed infants (human, not gorilla) to conduct his proposed experiment to produce lawyers, beggar-men, and thieves because he realized that not even he could create a crook out of thin air.

Heredity and Behavior

Let us consider some simple examples to show that genetics plays a role in the development of intelligence, personality, and emotional stability. In humans a condition known as phenylketonuria or PKU, inherited as an autosomal recessive, is characterized by phenylpyruvic acid in the urine and, if untreated, leads to severe mental retardation. Moreover, affected individuals are hyperactive, irritable, and subject to uncontrollable temper tantrums. Thus phenylketonuria, a simple recessive condition, influences the intelligence, the personality, and the emotional stability of affected persons.

Trisomy-21, the presence of an extra small autosome, leads to a condition known as mongolism or the Down syndrome. Afflicted with numerous physical abnormalities of the face, eyes, tongue, and other parts of the body, children with the Down syndrome often succumb to heart disease or infection. Like untreated PKU children, Down children are mentally retarded, but in contrast to them have warm, affectionate personalities. Faced with such examples, it is difficult to escape the conclusion that genes and chromosomes have something to do with the development of human intelligence, personality, and emotional stability.

A carping critic, however, may argue that such examples are irrelevant, that these conditions are highly aberrant and abnormal and that normal human intelligence and behavior are molded by environmental forces. Such arguments seem very like a modern-day version of Watson's Behaviorism. To argue that only extreme variants in intelligence and behavior are influenced by heredity but that variation within the normal range of intelligence and behavior is purely environmental is to ignore the evidence from numerous studies of variation in a wide range of species. No matter whether it is bristle number in flies, body size in mice, or yield in corn, numerous genes as well as the environment have been shown to be responsible for the phenotypic expression of these traits. Because behavior in animals, including humans, is dependent on the sensory, neural, motor, and other capacities of the organisms, all of which develop under the control of many different genes interacting with the environment, the inevitable conclusion is that a number of genes are involved in the development of behavior in humans and other animals.

It is clear that behavior in animals is the product of both genetic and environmental forces. A fertilized egg lacks behavior in the usual sense. Nonetheless, as an embryo grows and differentiates under the control of its genotype into an ever more complex organism with sense organs, nerves, muscles, and glands, its behavioral capacities increase and differentiate as well. A genotype, however, does not determine a particular phenotype; instead, it sets limits, a "range of reaction," on the range of possible phenotypes from that genotype. The particular phenotype that emerges is the product of the continuous interaction between the genes of the developing individual and its impinging environment. By the same token, the behavioral repertoire of a developing individual results from the interplay between its genes and the environment it encounters as it travels along the strip map of its life. To put it in simple terms, there is no instinct that is completely determined by heredity, just as there is no social or cultural or learned behavior that is purely environmental.

The Differences between the Sexes

Following this defense of the role of heredity in the development of behavior, a few words should be added to show that matters are not quite as simple as some would make them appear. It has been stated by Wilson (1978) in his Pulitzer Prize-winning book, *On Human Nature*, that men and women show genetic differences in behavior: ". . . modest genetic differences exist between the sexes; the behavioral genes interact with all existing environments to create a noticeable divergence in early psychological development . . ." and later, "The evidence for a genetic difference in behavior [between boys and girls] is varied and substantial" (p. 129). Although it is easy to agree that boys and girls and men and women differ in behavior, it is more difficult to agree that they differ simply because of differences in the behavioral genes they carry. The difficulty arises when one tries to figure out just where these behavioral genes might be located.

The human genome contains twenty-three pairs of chromosomes: twenty-two pairs of autosomes, which are the same in males and females, and one pair of sex chromosomes, which consist of two X chromosomes in females and an X and a Y chromosome in males. The autosomes and the X chromosomes are passed back and forth between males and females from generation to generation so that whatever genes they carry are shared by both sexes and can hardly be candidates for genes causing behavioral differences between men and women, at least not under any simple model. In humans, the Y chromosome is male-determining and is passed only from father to son. Therefore, there are just two genotypic differences between human males and females: males carry a small Y chromosome not found in females, and females have two doses of the X chromosome but males have just one.

Thus, of the forty-six chromosomes in human males and females, forty-five are alike in the loci they contain and thus in genetic content. The forty-sixth chromosome in females is a second X chromosome no different from the X chromosome carried by males. The forty-sixth chromosome in males is a small Y chromosome. If Wilson is right that there are genes governing the differences in behavior between boys and girls, then the Y chromosome is a candidate for carrying genes for male behavior, but where, then, do the genes for female behavior reside? The only remaining genotypic difference between males and females is the difference in dosage of the X chromosome. Obviously, something is amiss with this line of reasoning.

Another problem is that the small Y chromosome apparently carries relatively few genes. Furthermore, a segment of the Y seems to be homologous to part of the X. Thus it seems unlikely that the genes controlling the numerous characters associated with maleness are confined to the Y chromosome or that the characters associated with femaleness arise simply as the result of a double dose of the X chromosome in females.

One clue to the origin of the differences between the sexes in animals comes from the study of unusual types of sex determination and sexual differentiation. The mechanisms controlling sex determination and sexual differentiation are still far from being completely understood, and references in developmental biology should be consulted for additional information (e.g., Sang, 1984). In some reptiles such as turtles, incubation of the eggs at a high temperature (32°C) will produce a clutch all of one sex whereas incubation at a lower temperature (28°C or below) will produce a clutch of the opposite sex. Because the sex of the embryos is determined by a simple environmental cue, each fertilized egg must have a bisexual potentiality and must carry two sets of genes, one regulating male differentiation and the other female differentiation. The temperature of incubation determines which set of genes is called into play.

The situation in mammals is somewhat different, for in mammals the Y chromosome rather than temperature is male-determining. In humans, sexual development begins during the first six weeks after fertilization, but male and female embryos are outwardly indistinguishable. These early embryos have sexually indifferent gonads and both Müllerian and Wolffian ducts as well as external genital tubercles destined to become either the male penis or the female clitoris. However, the sex of these embryos can only be determined from their chromosomes.

The sex cells or germ cells do not originate in the gonads, but instead arise outside the body of the embryo in the margins of the yolk sac. From there they migrate to the gonads where they become the primordial germ cells, the precursors of either the sperm or the egg cells.

At about the seventh week, sexually indifferent human embryos begin sexual differentiation, which is completed by the end of the twelfth week. The gonads in males respond to the presence of the Y chromosome by secreting male hormones such as testosterone. The genes responsible for the production of the male hormones are located not on the Y chromosome but on the autosomes. These male hormones trigger a cascade of changes that lead to the differentiation of the male's primary and secondary sexual characteristics.

The indifferent paired gonads consist of an outer cortex and an inner medulla. During male sexual differentiation the medulla enlarges to form the testis but the cortex dwindles in importance. The Wolffian ducts, initially associated with primitive kidneys, differentiate into the epididymis, the vas deferens, and the seminal vesicles, but the Müllerian ducts degenerate.

In the absence of a Y chromosome, female sexual differentiation occurs. The outer cortex of the indifferent gonad differentiates into an ovary and the medullary sex cords regress. The Müllerian ducts differentiate into the Fallopian tubes, the uterus, and the upper vagina, but the Wolffian ducts degenerate.

Another bit of information bearing on the similarities and differences between males and females in mammals is that in the XX females, one or the other of the Xs is inactivated in every cell. Therefore, the active portion of the genome in both males and females consists of twenty-two pairs of autosomes and an X chromosome.

Sex determination and sexual differentiation take place differently in another carefully studied species, the fruit fly, *Drosophila melanogaster*. Here sex is determined by the balance between the X chromosomes and the autosomes. In *Drosophila*, a ratio of 1X to 2As produces a male; 2Xs to 2As results in a female. Both XY:AA and XO:AA give rise to a male phenotype whereas XX:AA and XXY:AA both look like females. Therefore, XXY:AA individuals look like females in *Drosophila* but like males in mammals, whereas an XO:AA individual is male-like in the flies but female-like in mammals. These and other chromosome combinations indicate that in mammals the Y chromosome is male-determining, but in *Drosophila* it is not. In the flies the primary role of the Y appears to be related to male fertility, for the otherwise phenotypically normal XO:AA males are sterile. Thus, phenotypic sex in *Drosophila* is determined, not by the Y, but by the balance between the X chromosomes and the autosomes.

A further noteworthy fact is that in *Drosophila*, dosage compensation occurs; that is, the amount of gene product from a single X in males is equivalent to the amount from two Xs in females. Therefore, mammals and fruit flies both have ways to compensate for the different doses of X chromosomes in males and females, mammals by X inactivation and flies by dosage compensation. This poses even more of a quandary, for it suggests that the active gene complements of males and females are even more similar than might have been suspected. Nevertheless, males and females do differ.

Still another type of evidence can shed light on the situation. When the normal sex-determining mechanisms go awry, abnormal sexual differentiation may lead to the formation of intersexes with intermediate sexual characteristics or else to aberrant hermaphrodites with both male and female traits. In other words, both male- and female-determining sets of genes must have been called into play to produce these sexual anomalies.

One of the less satisfactory aspects of our understanding of sexual differentiation in mammals is that each embryo appears to be preprogrammed to produce a female, but the production of male hormones under the influence of the Y chromosome diverts the course of development toward maleness. This leaves unexplained the origin of femaleness except as what develops in the absence of the Y chromosome. Nonetheless, the "dominance" of the male state was noted long ago in heterosexual cattle twins when anastomosis of their embryonic vascular systems led to masculinization of the female twin into a "freemartin" because male hormones and perhaps XY cells were transferred from the male to the female twin. The primacy of gonadal development in sexual differentiation is indicated by the fact that surgical removal of the gonads of small mammals during the indifferent period causes both sexes to develop as females.

The point of this digression into sex determination and sexual differentiation is to show that these phenomena are not the result of a simple genetic determinism such as that leading to albinism. Males do not carry one set of genes causing them to develop and act like males while females carry a different set causing them to develop and behave like females. It is much more subtle than that. The genetic constitution of males and females is demonstrably very similar. When the developmental pathway reaches the fork where the paths to the differentiation of males and females diverge, the future course of development is determined by a simple environmental or genetic cue that calls different arrays of genes into play. A form of biological determinism is certainly involved, but it is primarily a developmental determinism rather than a genetic determinism. The differences in behavior between males and females are not fixed in their genes, for both sexes carry essentially similar complements of genetic material. Instead, an initial triggering event precipitates a cascade of morphological, physiological, and behavioral changes that culminate in the formation of normal adult males or females.

The Biology of Behavior

Next, let us consider briefly the machinery that underlies behavior. Although extremely complex in some of the "higher" animals, in essence it is very simple, for it consists of just three main components: receptors, mediators, and effectors. The most familiar receptors are those for light, sound, touch, and for chemical molecules (taste and olfaction). These receptors provide sensory information to the organism so that it is kept aware of its environment and of any changes therein.

The mediators act as conductors and coordinators, for they transmit and process the information gathered by the receptors and coordinate the responses of the organism. In higher animals, the brain and nervous system are the primary mechanism for gathering information, processing it, and directing an appropriate response, although the endocrine system may also be intimately involved. The nervous system usually mediates rapid responses to the sensory input whereas the endocrine system's responses are slower but the effects last longer.

The responses are carried out by the effectors, which consist of muscles and glands. In the final analysis, all the complex and infinitely varied behaviors observed in higher animals result from muscular and glandular activity.

All the components of the machinery of behavior are the products of evolution. Sense organs, nervous system, glands, and muscles are all produced by historical and developmental processes. Although these components are often treated as separate traits and discussions of the evolution of the eye or the ear or the nervous system are common, this reductionist approach is quite misleading, for it seems to suggest that the eye or the ear or the nervous system has evolved independently of the rest of the organism. Even Darwin, trying to comprehend the evolution of the vertebrate eye, wrote as if it had evolved as a separate organ.

Selection and Behavior

The most vocal recent critics of modern evolutionary theory (e.g., Gould and Lewontin, 1979; Saunders, 1988) have attacked the theory because, they say, its advocates seek to explain the evolution of any trait in an organism by concocting "Just So" stories about its adaptive significance. If one story fails, it is quickly replaced by another. This criticism is certainly justified if an organism is truly thought to be just a bag of traits and evolution merely a series of changes in these traits. However, the critics' versions of the modern synthesis as nothing more than the natural selection of random variations in inherited traits (Ho and Fox, 1988; Saunders, 1988) or as "fundamentally a matter of selection for genes that make organisms" (Oyama, 1988) are so simplistic that they are little more than caricatures of modern evolutionary theory. The search for "new paradigms" on beyond Neo-Darwinism (Ho and Saunders, 1984; Ho and Fox, 1988) would rest on firmer ground if the seekers after a new paradigm used a more appropriate version of the old one.

It bears repeating that selection does not act on traits as such nor on genes as such. We may monitor evolutionary change by observing changes in traits or changes in gene frequencies, but that is not how evolution occurs. Natural selection acts on the phenotypes of individuals. There is a duality, a complexity to evolution that does not lend itself to glib explanations such as those offered earlier in this section. For example, to be biologically successful, an individual must both survive and reproduce. Hence, natural selection will favor those phenotypes that are able to do both. As the old song goes, you can't have one without the other. But even survival and reproduction are not enough for evolutionary change to occur. Only if the phenotypic differences between the more successful and the less successful individuals have an underlying genetic basis and are not merely the result of environmental and developmental quirks of fate can evolution occur. Therefore, natural selection is directly phenotypic, but indirectly genotypic. Genic selection, as such, does not occur, nor does the independent selection of traits.

Thus, direct phenotypic selection of individuals leads to indirect genotypic selection and ultimately to the changes in gene frequency that we study as indicators of evolutionary change. However, the individuals on whom selection acts are ephemeral, for they live and they die. Evolution does not occur in individuals, it occurs in a population. A population has continuity, an individual does not; nevertheless a population is made up of individuals. Here is another duality that takes some getting used to, for the population must be seen not just as a group of individuals but also as a pool of genes in the temporary custody of these individuals. This relationship is why evolution has so often been defined as changes in the kinds and frequencies of genes in populations even though that is not how it occurs.

The Role of the Environment

Another criticism of the modern synthesis is that it does not adequately recognize the role of the environment in evolution. Ho and Fox (1988) wrote, for example, "A predominant theme which runs through nearly all of this book [*Evolutionary Processes and Metaphors*] is an emphasis on the integration of the organism with nature." If, as seems to be the intent, "nature" is equated with the environment, this statement has its roots in the work of such authors as Henderson (1913), Waddington (1957), and Levins and Lewontin (1985). The central thesis is that organisms and environments are so intimately interrelated that they should not be considered separately. Levins and Lewontin (1985, p. 4) wrote of "the dialectical view of organism and environment as interpenetrating so that both are at the same time subjects and objects of the historical process"; and further (p. 99), "Just as there is no organism without an environment, so there is no environment without an organism." Such a statement is patent nonsense, for even though it is true that every organism occupies some sort of environment, it is not true that every environment contains some sort of organism. The surface of the moon comes to mind as an environment destitute of organisms, at least until the astronauts arrived, but they did not stay. Many environments such as the South Pole are so hostile that few if any organisms can exist there. The environment occupied by the passenger pigeon still exists, but the passenger pigeon is gone. Why then are such statements made?

Apparently, the purpose is to draw attention to the active role that animals play in choosing and even to some extent creating their own environments. Although plants are much less able to choose or to mold their environments, nonetheless Levins and Lewontin (1985, p. 53) specifically included them in their discussion. Furthermore, it is argued, the course of development in organisms is continuously influenced by the encroaching environment, and the organisms, in turn, constantly produce changes in the environment. These reciprocal interactions between an organism and its environment are the basis for the belief in the inseparability of organism and environment. There is also an echo here of the agonizing that has gone into the definition of the ecological niche, in particular whether a niche ever exists independent of an organism.

It is possible, however, to accept the idea that an organism is intimately associated with and influenced by its environment without accepting the opposite conclusion that the environment is equally sensitive to and influenced by the organism. The reason to hesitate over treating organisms and environments as equivalent and inseparable partners in the evolutionary process is very simple. Nothing in the physical environment plays a role equivalent to that of DNA in organisms. Although the environment may be modified by the presence of organisms, no DNA-like code capable of evolutionary change records the historical and developmental properties of the environment. To suggest otherwise seems like a new form of vitalism, conferring on the environment the properties of living organisms. There is no intent here to denigrate the role of the environment in evolution; the intent is simply to suggest that environments and organisms do not play equivalent evolutionary roles. Both are important, but they are obviously different.

Ironically, the same authors who criticized the modern synthesis for minimizing the importance of the environment also attacked the "adaptationist programme," which is the ultimate in environmental theories of evolution, based as it is on natural selection for functional adaptation to existing or changing environmental conditions.

The Evolution of the Eye

Let us return to the evolution of the eye. Although vertebrate eyes are distinct organs, they do not evolve separately and independently of the organism. The eyes are an integral part of the phenotype of an animal, and the success or failure of the eyes as a trait will depend, not on their functioning as eyes, per se, but on the extent to which they contribute to the survival and reproduction of the organisms possessing them. If there is genetic variation leading to phenotypic variation among the eyes of different individuals, and if this phenotypic variation in turn permits some individuals to survive and repro-

duce more successfully than others, the gene pool and the phenotypes of the eyes in the population will change through time. These
changes occur, not to produce better eyes, but to produce better organisms, organisms better able to survive and reproduce. The recent
discussions about the units or levels or hierarchies of selection, ranging from molecules through organelles, cells, organs, and individuals
to kin, groups, species, and communities, obscure the simple fact that
individual selection is the primary selective force bringing about evolutionary change. Although evolution can be studied at these other
levels, the ultimate determining factor for the course of evolution is
the success or failure of individuals.

Moreover, the very use of the concept "levels of selection" leads
almost automatically to a mind-set that equates evolution with natural
selection, as if the complexity of evolutionary change is tied to the
complexity of a hierarchy of selection pressures. The evolution of the
eyes in cave animals alone may help to disabuse us of this notion.
Cave animals, which spend their entire lives in total darkness, are ordinarily descended from species that lived above ground in the vicin-
ity of the caves. Large caves are characterized not only by the absence
of light but by relatively constant temperatures and by rather high
and constant humidities. Food supplies are limited because, in the absence of photosynthesis, nearly all food must be imported from outside the caves. Two of the more striking features about cave animals
are the loss of pigment and the loss or degeneration of the eyes.

At least a dozen theories have been proposed to account for regressive evolution of this sort (Barr, 1968). The simplest is pure Lamarckism: because an animal no longer uses its eyes, they degenerate, and
the degeneration is inherited. In the dark, an animal no longer needs
protective coloration, so it is lost. Unfortunately, the evidence for the
inheritance of acquired characteristics based on use and disuse or the
needs of the organism has never withstood close scrutiny. Experimental results seeming to support Lamarckism have been reported
sporadically over the years, but either the experiments could not be
replicated or other more plausible explanations were available.

Another theory is that cave animals have lost their eyes because of
positive selection against eyes in the cave habitat. It is argued that the
energy supply in caves is so limited that selection will eliminate organs no longer useful to the animal in order to conserve energy. This
adaptationist "Just So" story may seem plausible for cavernicolous
species, but it fails for parasites.

Regressive Evolution

When compared to their free-living relatives, internal parasites often show regressive evolution in their sense organs and in their nervous, digestive, and other systems. But even though their habitat is in many ways analogous to a cave, one major difference is that internal parasites are surrounded by an abundance of food. Therefore, selection for energy conservation, offered as an explanation for regressive evolution in cavernicolous animals, lacks generality, for it can hardly account for the rudimentation of organs in internal parasites.

How else might regressive evolution occur? Presumably, the highly efficient eyes of higher animals evolved from the primitive light receptors of lower forms because each advance conferred an adaptive advantage. Natural selection through eons of time favored those individuals with heritable variations providing, in one way or another, more effective light reception under the prevailing conditions. What happens when such animals take up residence in caves, living in total darkness?

First, light receptors will no longer be adaptively relevant. It is sometimes said that the eyes degenerate and pigment is lost in cave animals because selection is relaxed. (See the discussions in Heuts [1953], Barr [1968], and Dobzhansky [1970].) This statement is so misleading that it is better left unsaid. The concept of relaxed selection was discussed earlier in relation to population expansion. Here again it emerges, and again it is of dubious value. Selection pressures on previously free-living animals that take up life as troglobites or as internal parasites are not really relaxed; they are different. If we learn nothing else from population genetics, it should be that evolutionary change does not occur unless one or more of the four factors— selection, mutation, migration, and drift—brings about shifts in the kinds or frequencies of genes. Therefore, the mere "relaxation" of selection is not going to lead to regressive evolution in troglobites or rudimentation of structures in parasites. A relaxation of selection suggests the absence of selection, yet no evolutionary change can occur unless one or more of the four evolutionary factors is at work.

How then can regressive evolution be explained? So far we have discussed Lamarckism, selection for energy efficiency, and relaxed selection as possible explanations for regressive evolution. Another possibility is differential migration, with the eyeless, less pigmented forms being photophobic and tending to hide in caves whereas the eyed, pigmented forms are photophilic and continue to live above ground. This theory has several versions but none explains the loss of eyes and pigment; rather they suggest how the two types might sort themselves out once they have originated.

Another type of theory dealing more directly with the origin of the cavernicolous types is that loss of the eyes and pigment results from the accumulation of random mutations (Brace 1963), or from mutation pressure. The essential idea here is that most mutations tend to have detrimental rather than favorable effects. So long as the structures or organs affected by the mutations play a functional role in the organism, these harmful mutations will be eliminated by natural selection. If the function is lost, as, for example, protective coloration and eyes become useless in total darkness, selection will no longer maintain the integrity of the developmental systems that give rise to eyes or to cryptic coloration. Because the effects of these previously harmful mutations are no longer opposed by natural selection, they will tend to accumulate and regressive evolution will result, not from relaxed selection, but from mutation pressure. However, Wright (1964) and Prout (1964) both argued that mutation pressure alone is unlikely to be the cause of regressive evolution. To reach this conclusion, they had to make certain assumptions about mutation rates and the numbers of loci involved. Whether the conclusion would have held up under other assumptions (e.g., higher mutation rates at more loci) is not clear. In any case, their argument was not that mutation had no role to play but rather that it was not the only factor involved in regressive evolution. Prout suggested that multiple causes, including selection, pleiotropy, drift, mutation, and even meiotic drive, were apt to be responsible for structural reduction. For example, an unused eye, rather than being adaptively neutral or merely a waste of energy, may actually be harmful if it is subject to injury or infection and thus may be eliminated by selection. Wright favored a version of his shifting balance theory of evolution with emphasis on selection for the pleiotropic effects of newly favored mutant alleles that are replacing the existing alleles of the free-living or epigean populations. Wright's ideas are appealing because they provide a possible explanation for the persistence of vestigial structures such as the little toe in guinea pigs or the imbedded pelvic girdle of pythons and whales.

Given the diversity and large numbers of cavernicolous and parasitic species, it seems highly improbable that the regressive evolution so often observed in these species has the same explanation in all cases. We have already seen that selection for energy efficiency, a possible explanation for regression in cave animals, is not plausible for parasites. Let us now consider in turn the role of each of the four evolutionary factors in regressive evolution. Mutations, whose rates may be influenced by environmental conditions but are not ordinarily thought to be adaptive responses to those conditions, will continue to occur in cavernicolous and parasitic as well as free-living populations. Mutation pressure alone is unlikely to determine the fate of these mutations because it acts so slowly.

When a population invades a new habitat, whether a cave or an organism, different selection pressures will be brought to bear on it. Because every species inhabits some kind of environment, it is subjected to certain kinds of selection pressures, and when the habitat changes, the selective forces inevitably change. Given the dramatic change in mode of life from free-living to parasitic or from epigean to cavernicolous, it seems highly probable that opportunistic evolution will come into play, that any genetic variant, whether of large or small effect, that enables an organism to survive and reproduce in this new habitat will be favored by selection. Furthermore, the adaptive shifts under these circumstances may occur quite rapidly, either through selection for mutations of major effect or for favorable combinations of polygenes, or most likely, some combination of both. Once the adaptive shift has been made, selection will continue to modify and refine the adaptation of the organisms to their new way of life.

The role of random genetic drift and of migration in the evolution of parasites and cave animals is more conjectural. It seems probable that the effective size of the breeding populations in both types of species, given the nature of their life histories, may often be quite small, in which case drift may well be a factor in these populations, particularly because it may lead to genetic divergence among them. If photophilic and photophobic types show differential migration, which is actually a form of natural selection, their isolation and genetic divergence from one another would be enhanced. On the other hand, if migration and gene flow continue between free-living and parasitic or cavernicolous forms, the evolution of parasites and troglobites would be greatly slowed or even blocked entirely. Perhaps it should be added that it has been implicitly assumed, as seems logical, that both parasites and cave animals have originated from free-living epigean species.

What, finally, can be said about the mechanism of regressive evolution? Mutation is the ultimate source of the variants that lead to the loss or degeneration of structures, but mutation pressure alone seems unlikely to account for rudimentation. Changed selection pressures must also have played a role. Although the sense organs and the nervous, digestive, muscular, circulatory, and excretory systems have often regressed in internal parasites, their reproductive systems have usually become more complex and hypertrophied compared to those of their free-living relatives. The whole genetic system as well as the biochemistry, physiology, morphology, and behavior of these species has undergone change, in some cases revolutionary change, as these organisms have adapted to new modes of life. These transitions could occur quite rapidly under the combined impact of mutation, selection, and drift, but gene flow seems unlikely to have played a major role. The course of events in any particular case can best be studied by comparing a parasitic or cavernicolous species to its nearest free-living relatives. Some of the molecular and genetic techniques now available may provide new insights into the evolution of such species.

Most evolutionary studies have been devoted to the evolution of ever-increasing complexity or to progressive evolution, as it is sometimes called, with relatively little attention paid to regressive evolution. Rensch (1960), for example, devoted an entire chapter to "anagenesis" or progressive evolution but very little space to regressive evolution. Price (1980) wrote a work entitled "Evolutionary Biology of Parasites" but did not discuss regressive evolution or the origin of parasitism at all. He did, however, present some data on the number of known parasitic species as compared to free-living species in order to demonstrate the preponderance of parasites. There is, however, a simpler and more convincing way to illustrate this point. Every known free-living species supports at least one parasitic species and usually many more. If one includes viruses and bacteria with the parasites treated in such texts as Schmidt and Roberts (1985), it seems safe to conclude that the number of parasitic species must be at least ten times greater than the number of free-living species. In fact, if allowance is made for the numbers of parasitic species yet to be discovered, there are probably at least one hundred times more parasites than free-living species. Nevertheless, the evolution of free-living species of animals and plants has drawn much more attention than the evolution of parasites. Since the overwhelming majority of species is parasitic, such emphasis seems misplaced. More intensive study of the evolution of parasites and of parasitism could bring a whole new dimension to the study of evolution.

For one thing, "regressive" evolution would be seen as a phenomenon equal in interest and importance to "progressive" evolution. For another, many more cases of sympatric speciation would be uncovered. Moreover, the evolution of parasitism should provide a fertile field for the discovery of more cases of saltational evolution. Although parasites often show regressive evolution in many aspects of their form and function, their life histories are often far more complex than those of their nonparasitic relatives. The opportunities for new discoveries about the origin and evolution and genetics of the intimate association between a host and its parasites seem endless.

The Biological Environment

Parasitism is an instance where one organism becomes the environment of another. Here the phenotype and genotype of the parasite must be adapted to the environment provided by the phenotype and genotype of the host, and the phenotype and genotype of the host must permit it to withstand the invasion of the parasites. Here is the perfect example of Levins and Lewontin's (1985) "dialectical view of organism and environment as interpenetrating so that both are at the same time subjects and objects of the historical process." However, it is quite clear (pp. 51–58) that host-parasite or other interspecific relationships were not at all what Levins and Lewontin had in mind. Like most evolutionary biologists, they thought primarily in terms of a single free-living species and its relationship with and impact on its physical environment. They pointed out that the environment is constantly changing because organisms deplete resources, excrete waste products, and modify the oxygen, carbon dioxide, and water content as well as the temperature, light, and substrate of the environment.

The environment is not just physical, however; it is also biological. All species are bound together in an intricate web of primary producers, herbivores, omnivores, carnivores, parasites, saprophages, saprophytes, and decomposers. Even autotrophs such as green plants, which in theory are able to live independently of other species (unless they require insect pollination), seldom do so, for they may be eaten by herbivores or attacked by parasites. The biological environment of any individual is further complicated by such interspecific relationships as competition, commensalism, and mutualism as well as by many types of interactions with members of its own species.

This intricate web of associations is a major factor in determining the course of evolution. All species must adapt to their biological as well as to their physical environment, and their evolution is influenced and shaped by their adaptive responses to the presence and behavior of members of their own and many other species. The importance of the biological environment to evolution has been recognized since the time of Darwin, for example, in his discussions of intra- and interspecific competition (1859) and the theory of sexual selection (1871). More recently, the evolutionary relationship between predator and prey, parasite and host, or herbivore and plant has been called an "arms race" by Dawkins and Krebs (1979). It had earlier been dubbed the "Red Queen hypothesis" (Van Valen, 1973) because both species, like Alice's Red Queen, have to evolve as fast as they can just to keep their places in the evolutionary race. The subject has also been treated under the heading "coevolution" in several books (Thompson, 1982; Futuyma and Slatkin, 1983; Nitecki, 1983).

In all of these cases, each species is both subject and object of the historical process. The gene pool of each evolves in response to the presence of other species. Flor (1956) was the first to describe the exact relationship between the genotypes of host and parasite in his studies of flax and flax rust. He discovered a gene-for-gene relation between the resistance of cultivated flax varieties and the ability to overcome resistance by strains of the flax-rust fungus. This pioneer work suggested how the continual interplay between the genotypes of different species can lead to their continued evolution.

The physical environment also changes as a result of the presence of living organisms, but it does not evolve, for it has no genetic system capable of evolution. Fisher (1930) was one of the first to suggest that the environment is "constantly deteriorating" and that organisms must continually evolve to adapt to changing physical and biological conditions. The use of "deteriorating" in this context seems misleading. The deterioration of a freshly yeasted cornmeal-agar medium as a suitable habitat for *Drosophila melanogaster* can be easily observed in the lab, but this same medium, as it ages and putrefies, becomes an increasingly favorable habitat for another fly, *Drosophila funebris* (Merrell, 1951). Certainly any environment is always changing, but the earth, which constantly gains energy from the sun, is not a closed system, and therefore the environment does not really deteriorate; it changes. One might just as well argue that because of ongoing arms races, species are always "improving." However, it is as difficult to define "evolutionary progress" as it is to define "environmental deterioration," and both expressions are relative rather than absolute.

The pitfalls lurking beneath the words we use can be illustrated by "parasitism," which is often regarded as a sign of degeneracy or of a morally and ethically reprehensible way of life. Synonyms for parasite range from sycophant, hanger-on, toady, leech, and sponger to panhandler, cadger, and flunky. All are derogatory. For this reason, many are troubled by the thought that parasitic species far outnumber the free-living, upright, decent species of the world. The truth, of course, is that all animal species are dependent on other species for their very existence. Parasites are simply more efficient than other kinds of organisms at exploiting other species, so perhaps they should be considered the epitome of evolution rather than examples of degeneracy. If this line of thought is carried too far, it could end up in a quagmire of terminology, but it is worth noting because all evolutionary change in parasites is not regressive or degenerative. Parasites manifest some of the most complex life histories, developmental sequences, and genetic systems known.

The title of this chapter is "Behavior and Evolution" and we seem to have wandered rather far afield from the subject. In fact, however, we have not, for the intent has been to broaden our concept of behavior to include more than the usual types of behavior studied by ethologists. The biological environment of an organism consists not just of other members of its own species and of competing species, but of all the species with which it comes in contact. Each species has a limited range of physical conditions (temperature, moisture, light, oxygen, and so on) in which it can survive. These conditions bring one set of selection pressures to bear on an organism. The many species with which an organism coexists and interacts represent another set of selection pressures. Some may be mates, some may be food, others may be predators, parasites, hosts, competitors, or mutualists. Furthermore, all of these physical and biological factors (i.e., all of these selection pressures) may impinge on the organism at the same time. The way the organism behaves in response to all of these factors will determine whether it is able to survive and reproduce. Clearly, behavior creates many of the waves on the adaptive seascape and influences the course of evolution in diverse ways.

Most discussions of adaptation have concentrated on the adaptive significance of particular traits, which has properly been criticized because that is not how adaptation occurs, or on the adaptation of a population to its "environment," with emphasis on the physical rather than the biological characteristics of the environment. Recently, the possible role of biological interactions in evolution has been recognized in the Red Queen hypothesis, arms races, and coevolution. Most of these discussions, however, have dealt with the interactions between just two species (i.e., predator and prey, plant and herbivore, parasite and host), and they have not been framed in behavioral terms. The reality is that a population must adapt, not just to one other species, but to its physical surroundings and simultaneously to all of the species that affect its existence, whether they be mates, prey, parasites, hosts, competitors, or merely the trees where they perch. The association between a parasite and its host is more intimate than most other relationships, but even parasites, in the course of their life cycle, are exposed at times to physical and biological conditions quite different from their cozy abode within their hosts.

Approaches to the Study of Evolution

Saunders (1988, p. 278) and others have claimed that the modern synthetic theory of evolution is not falsifiable and therefore is not really a scientific theory at all. If his version of Neo-Darwinism were correct, this might be true, for he wrote (p. 275): "Neo-Darwinist explanation consists largely of suggesting selective advantages for various traits of organisms. Because the theory is unfalsifiable, with a little ingenuity such an explanation can always be found." He also stated (p. 276): ". . . what it [Neo-Darwinism] does explain are only some of the minutiae of evolution rather than the major features: changes in the coloration of moths, but not how there came to be moths in the first place." This latter statement evokes the shades of Goldschmidt (microevolution and macroevolution, 1940) and of the biometricians and developmental biologists of the early days of this century, who felt that Mendelian genetics could only account for the more superficial or trivial traits of organisms but not for their more fundamental characteristics. Given the many advances in genetics since those days and the failure to find any alternative mechanism of heredity for the more important traits, Saunders's statement sounds oddly anachronistic.

Let us deal first with the idea that Neo-Darwinism is unfalsifiable. If all evolutionary biologists did was sit around concocting "Just So" stories about adaptive traits, as Saunders and many others seem to think, then Neo-Darwinism would indeed be unfalsifiable. Hailman (1988) suggested that the two principal methods for identifying adaptations (which he preferred to call "suitabilities") were optimality theory and the comparative method. The optimality method attempts to demonstrate that a given species "is optimally constructed (or behaves optimally) relative to its environmental problems." The comparative method "seeks to establish the existence of character-environment correlations among unrelated species." Hailman concluded that optimality theory failed as a way to identify adaptations but was optimistic about the comparative method.

Optimality theory has probably been most widely used in relation to optimal foraging strategy (OFS) (Kamil and Sargent, 1981; Kamil, Krebs, and Pulliam, 1987; Gray, 1988). The essence of the concept is that an animal employs the most "efficient" or optimal strategy when foraging for food and that this adaptive strategy has evolved through natural selection. Some studies of OFS have reported finding one but others have not (Gray, 1987). Quite apart from the difficulty of determining just what is optimal or the most efficient way to forage, there is a more fundamental difficulty. Foraging is a behavioral trait rather than a morphological trait, but natural selection does not act on traits; it acts on the phenotypes of individuals. Optimal foraging strategy is simply another "Just So" story about behavior instead of morphology.

If organisms had one-track minds that led them only to food, studies of optimal foraging strategy might be more rewarding. Important though food may be, however, it is not the only factor affecting an animal's behavior. At the same time it is seeking food, it may also be seeking a mate, confronting a competitor, evading a predator, and so on. The actual behavior of the animal at any given time can be thought of as the vector resulting from all these contending forces. Optimal foraging theory can provide, at best, only a distorted, one-dimensional view of an animal's behavior.

The comparative method favored by Hailman (1988) depends on "the comparison of species in at least two independent taxa such that some members of each taxon inhabit the same environment as some members of the other taxon and each taxon also contains other species inhabiting different environments. Only when unrelated species living in the same environment share characteristics (convergent evolution) and differ in these characteristics from their relatives living in other environments, is the reality of evolutionary suitabilities [adaptations] established." These specifications are quite explicit although just what is meant by "two independent taxa" or "unrelated species" is not stated. However, exactly the same conditions (sympatric and allopatric populations of two different taxa) are needed to study character displacement (evolutionary divergence thought to result from competition between sympatric species). When convergence is found, Hailman attributes it to similar adaptive responses by different species to a common environment. The comparative method fails, on the other hand, when sympatric taxa diverge compared to allopatric taxa. In this case, divergence is attributed to competition between species with overlapping niches, for it cannot be explained as adaptation to a common environment.

The comparative method has other problems as well. One is that the choice of characters to match with environments is subjective. Moreover, all that is established are correlations between traits and environments, which tell us little or nothing about causation. Furthermore, this approach has the same flaw as many others, for it deals with the evolution of traits, and by now it should be evident that traits, as such, do not evolve.

As noted earlier, Saunders (1988) and others consider modern evolutionary theory to be unfalsifiable and therefore not really a scientific theory at all. However, science is distinguished from other areas of study such as philosophy, religion, and the arts because its body of knowledge and principles are gained through observation, study, and experiment. Strangely lacking from much of the evolutionary literature is any suggestion that the way to test the validity of evolutionary theory is through experimentation. An experimental approach to the study of evolution poses many problems, but the potential rewards are far greater than can be won with optimality theory or the comparative method. Because many evolutionists are not experimentalists, they apparently consider experimental tests of evolutionary theories either impossible or impractical, but that is not necessarily the case.

For example, the modern synthesis predicts that no evolutionary change is possible in a population lacking genetic variability. At a time when virtually all entomologists believed that DDT resistance resulted from physiological adaptation rather than genetic adaptation, it seemed desirable to try to demonstrate that DDT resistance would result from natural selection in genetically variable populations but that homozygous populations would not respond to selection. The results of selection experiments with highly inbred stocks and with more heterozygous laboratory and wild-derived stocks of *Drosophila melanogaster* confirmed these predictions. The inbred stocks showed no increase in DDT resistance during thirty months of selection, but the more heterozygous laboratory and wild stocks developed varying degrees of increased resistance (Merrell and Underhill, 1956b). The homozygous inbred flies were never able to adapt to DDT whereas after the resistant strains developed, previously unexposed flies from these strains were resistant when first exposed to DDT, clear evidence that their adaptation to DDT was genetic and not physiological. The DDT-resistant stocks and their corresponding susceptible controls diverged not only in DDT resistance but in most cases in fecundity, fertility, and longevity as well (Underhill and Merrell, 1966). By any definition, these experiments resulted in adaptive evolutionary changes brought about by natural selection.

Some years ago in Florida I saw pigeons perching on telephone wires. The common pigeon is descended from the European wild pigeon or rock dove (*Columba livia*) from which most domestic varieties are derived. Following their introduction into North America, they became widespread and abundant, and I had previously seen them perching in all sorts of places, but never on telephone wires. Here was a simple behavioral difference between Florida pigeons and pigeons elsewhere. Like a good Neo-Darwinian, I immediately concocted a "Just So" story to account for the difference: because Florida can be extremely hot, the pigeons had evolved this behavior to avoid perching on the hot roofs and ledges where they normally perch. In the absence of any supporting evidence, this plausible hypothesis falls into the realm of speculation of the sort that has given the adaptive paradigm and Neo-Darwinism a bad name. It is also possible, for instance, that this behavior arose by chance, is learned by the young from their parents, and has no evolutionary significance whatever.

How could the causes of this behavioral difference be investigated experimentally? A first step would be to bring wire-perching and non-wire-perching pigeons together in a common environment to see whether the behavioral difference persists. A second step would be to rear offspring of the wire-perching Florida birds in a cooler climate to see whether they would still perch on wires as their parents did. Cross-fostering experiments could be used to test whether or not perching sites are learned by imitation of the parents. If any of these experiments indicate a genetic component to this behavioral difference, the next step would be to cross wire-perching and non-wire-perching birds to initiate a study of the genetics of the trait.

This sort of experimental approach could reveal whether the behavior has a genetic component but not whether it has adaptive significance. To test for that, it would be necessary to determine the relative fitness of the two types of birds in the equivalent of reciprocal transplant experiments or else devise competition experiments between them in at least two environments. Obviously the study of the adaptive significance of even such a simple behavioral difference as this soon poses formidable experimental challenges. "Just So" stories are far easier to concoct than good experiments. Little wonder that students of adaptation usually use indirect approaches rather than experiments to study adaptation. However, the experimental approach, though difficult, is not impossible and is the next necessary and logical step in the study of adaptation, one of the most fundamental problems in biology.

It should be noted that perching on telephone wires is a behavioral trait, that we are back to discussing the adaptive value of traits. Practical considerations in devising experiments almost dictate this approach. The behavioral difference is well defined and easy to score, the genetics should be relatively simple, and the environments favoring such differences in behavior should be easy to specify. The inherent complexities in the study of adaptation are so great that it is desirable to simplify the task whenever possible. The risk in studying the adaptive value of a trait lies in regarding it as independent of the rest of the phenotype and evolving in a vacuum when in fact it may be only a minor component in the fitness of the organisms. This approach, however, provides an entrée into the study of adaptation and, because it requires close scrutiny of the autecology of the organisms, it should lead to further insights and experiments and ultimately to a better understanding of the ways in which organisms are able to ride the waves of the adaptive seascape.

Chapter 14
Where Are We Now?

Modern evolutionary theory has certainly progressed on beyond Darwinism. Darwinism has acquired many different meanings (Mayr, 1991), but evolution by natural selection, its most common present-day definition, is only a partial and incomplete statement of the modern theory of the mechanism of evolution. Darwin's theory was incomplete primarily because he lacked adequate knowledge of heredity. The mechanism of evolution is a genetic mechanism, for evolutionary change results from genetic changes in populations. Pangenesis, Darwin's theory of heredity, and his belief in Lamarckian inheritance of acquired characters were soon shown to be wrong. His greatest contribution to our understanding of evolution was to show that the mechanism for adaptation is natural selection, thus solving one of the great puzzles in biology.

The construction of the genetic underpinnings to the modern theory of evolution began with the rediscovery of Mendelism in 1900. Knowledge of genetics has grown at an accelerating pace ever since, culminating with the advances in molecular genetics. The framework for the modern theory of the mechanism of evolution was constructed primarily by the population geneticists Chetverikov, Fisher, Haldane, and Wright. Their contribution was to incorporate mutation, gene flow, and random genetic drift as well as selection into the mechanism of evolution, introducing greater complexity and stochasticity into the modern synthesis. The addition of these factors plus a knowledge of genetics make present-day understanding of how evolution occurs far more advanced than it was in Darwin's day. Some have argued that it is time to seek a new paradigm, to go on beyond the modern synthesis, but the modern synthesis is far more resilient and adaptable than is sometimes recognized, and as yet there is no need to replace it.

The concept of punctuated equilibria, for example, introduced by Eldredge and Gould (1972) as an alternative to the modern synthesis, was readily accommodated within the existing framework. The initial turbulence created by this supposedly new concept has subsided, and we have now progressed on beyond punctuated equilibria.

The tremendous advances in our understanding of heredity resulting from research in molecular genetics seemed likely to require modification of the modern synthesis. For instance, the discovery of high levels of genetic heterozygosity and genetic polymorphism in natural populations not only provided better knowledge of the genetic composition of such populations but led to conceptual innovations. Three concepts drew considerable strength from these findings: genetic load, neutralism, and the molecular clock.

Although the concept of genetic load stimulated a great deal of worthwhile discussion and research, in the final analysis it turned out to be of limited usefulness and in some ways misleading in its stress on the cost of evolution rather than the cost of not evolving. Again, we seem to have passed on beyond genetic load.

The concept of a molecular clock has sometimes seemed like a will-o'-the-wisp, for the more carefully the data were scrutinized, the more elusive the clock became. The idea that evolution chugs along at a rate dependent on the rate of substitution of mutations seems simply at odds with all that is known about evolution and also with what has been learned from detailed molecular studies. Gillespie (1991, p. 140) has even gone so far as to state that "the clock does not exist."

Similarly, the bold new neutral theory of molecular evolution has undergone modifications and revisions since it was first proposed. Alleles are now referred to as nearly neutral rather than neutral, and as pointed out earlier (Merrell, 1981, p. 326), ". . . there is all the difference in the world between *not pregnant* and *slightly pregnant* just as there is between *neutral* and *nearly neutral*." Furthermore, after an extensive analysis, Gillespie (1991, p. 291) concluded that natural selection is "a major force contributing to the evolution of proteins" and that "most amino acid substitutions are adaptive." Thus, after being hung up on these topics for some time, we now seem to have moved on beyond genetic load, the molecular clock, and neutralism, but the modern synthesis survives.

When genetics became a science in 1900, one of the first questions was whether Mendel's laws of segregation and independent assortment applied to other species of plants and animals as well as to garden peas. Although we now take it for granted, it is a great blessing that the chromosomal mechanism of heredity turned out to be universal, for it greatly simplified the study of genetics to know that findings in one species were apt to be universally applicable. Similarly, when the molecular nature of the gene was studied, a universal genetic code was revealed; the same code was used by species of plants, animals, and micro-organisms to transmit, transcribe, and translate hereditary information. The major advances in genetics were based on the study of a rather small number of species such as peas, *Drosophila*, corn, mice, and more recently, fungi such as *Neurospora* and certain yeasts, the bacterium *Escherichia coli*, a few strains of bacterial viruses, and the roundworm *Caenorhabditis elegans*. This small, motley crew of species has provided the basis for our knowledge of a genetic mechanism that has turned out to be universal, for it operates in nearly all species of organisms. Imagine how different the science of genetics would be if the mechanism of heredity were not universal but differed from one group of species to another.

When we turn to the mechanism of evolution, again we seem to find a universal mechanism, the modern synthesis, involving mutation, selection, migration, and drift. However, these factors provide the framework within which evolution occurs and are at least one step removed from the underlying universal genetic mechanism. The actual course of evolution in any species is determined by the interplay among all these factors and the biological and physical environment. Although geographical or allopatric speciation involving gradual evolution by means of numerous small genetic changes has been promoted as the universal mode of evolution, it has become increasingly clear that there is no universal mode of evolution comparable to the universal mechanism of heredity. Instead, within the framework provided by the modern synthesis, species have originated in a variety of ways, and the patterns and modes of evolution are diverse.

One reason that geographical speciation has been so popular is that evolutionary studies have concentrated on relatively limited groups of organisms: the vertebrates, particularly birds and mammals, the higher plants, and a few insect taxa. Even within this biased sample of species, however, it is already clear that speciation may be saltational as well as gradual and sympatric as well as allopatric. As evolutionary studies are broadened to include a wider range of species among lower plants, invertebrates, and micro-organisms, a new and richer variety of modes of evolution seems likely to emerge. Given the enormous numbers of parasitic species, the study of the evolution of parasites and parasitism seems likely to be particularly rewarding. If the two great problems of evolution are adaptation and speciation (Waddington, 1968), it seems safe to say that we have a better understanding of the mechanism of adaptation at present than we do of the mechanisms of speciation.

Rather than seek a universal mode of evolution, the evidence suggests that we should use a pluralistic approach to study the origin and evolution of species. Just as no single theory adequately explains all the phenomena subsumed under the label of dominance, no single mode of evolution can account for the variety of ways that species have evolved. Thus, even though there is overwhelming evidence for adaptive evolution, there is also considerable evidence for nonadaptive evolution. Moreover, speciation may be sympatric as well as allopatric and saltational as well as gradual. Furthermore, the two-stage theory of the role of genes in evolution permits oligogenes, particularly dominant oligogenes, to assume their proper role along with polygenes as agents of evolutionary change.

The adaptive landscape has been the dominant metaphor in evolutionary biology for some sixty years. The concept of adaptive peaks and valleys has had considerable heuristic appeal, but an adaptive landscape also sets a metaphorical trap, for it carries the implication of a more or less fixed topographical map of fitness. Well-adapted organisms occupy the adaptive peaks, but there they are trapped, for in order to evolve, they must somehow move to other peaks, in the process moving through valleys of lower fitness.

How to move from one fixed adaptive peak to another has been an ongoing problem in evolutionary biology for decades. One way to escape from this trap is to change the metaphor. The adaptive seascape provides a far more realistic analogue of the ever-changing physical and biological environment of organisms than the adaptive landscape. The adaptive landscape leads to the concept of a dynamic gene pool evolving under constant selection pressures in a static environment. The adaptive seascape, on the other hand, emphasizes that environmental variation as well as genetic variation comes into play in the dynamic interactions leading to the origin of species. Furthermore, it eliminates the metaphorical trap that confines well-adapted populations to adaptive peaks, for in an adaptive seascape there are no fixed adaptive peaks.

The other great advantage of the adaptive seascape is that it draws attention to the variable environment as an instrument of evolutionary change equivalent to the variable gene pool of the evolving species. Because the biological environment is such a significant part of this variable environment, the evolutionary impact of behavior, both individual and social, will inevitably become a focal point in evolutionary studies. The best prospect for advancing our knowledge of the ways species respond and adapt to their changing biological and physical environment—in other words, for studying how the mechanism of evolution actually works—lies in the study of the autecology and the ecological genetics of individual species. As the study broadens to encompass not just the members of the species itself, but also its competitors, predators, parasites, food, and habitat, the complexity of the evolutionary process will become manifest.

Quite apart from the intellectual challenge of trying to solve the central problem of biology, the origin and evolution of species, evolutionary studies have immediate and practical benefits. Plant and animal breeders, for example, are practical evolutionists, for they try to produce evolutionary changes in their breeding populations through artificial selection toward more economically desirable varieties. At the same time they are constantly confronted by the effects of natural selection, for their artificially selected populations are also undergoing natural selection in order to survive and reproduce in the environment where they are being raised. Plant breeders often become involved in an evolutionary footrace, for almost as soon as they develop a rust-resistant strain of wheat, for example, a mutant strain of rust appears capable of attacking the formerly resistant wheat, and the wheat breeders must try to keep their wheat varieties one step ahead of the evolving rusts.

Microbiologists face similar problems. The medical use of antibiotics and drugs has caused populations of bacteria to evolve resistance to these agents. A longitudinal study in a hospital revealed, for example, that over the course of a decade, the bacterial populations had evolved an ever-changing spectrum of resistant strains. Higher organisms have also shown a distressing ability to evolve resistance to insecticides, pesticides, and herbicides. Although evolution is often associated with the age of the dinosaurs, in truth it is going on around us all the time.

At present there is considerable concern about the loss of species diversity through the destruction of habitats such as the tropical rain forest. Such losses are cause for serious concern for many reasons, not least of which is aesthetic, and have led to many efforts toward the preservation of pristine natural areas and parks where existing species can continue to survive. Conservation biologists have also been involved in attempts to save threatened and endangered species from extinction, often through captive breeding programs as well as habitat preservation. All such efforts plus the attempts to substitute biological pest control for the use of pesticides qualify as practical exercises in evolutionary biology.

In the discussions about the loss of species diversity there is seldom any mention of the fact that unless all life becomes extinct, evolution will not cease and new species will continue to evolve. Much of the present destabilization of the earth and its inhabitants results from the activities of the ever-increasing numbers of humans. Human population pressure is responsible for most of the destruction of habitat and extinction of species. One of the more striking biological phenomena resulting from human activity is the widespread dispersal over the earth of species formerly restricted to limited geographical areas.

Some of the greatest present opportunities to study evolution in action lie in the study of introduced species living in new and different and often disturbed habitats. Although most studies of natural populations now tend to concentrate on species living in wildlife refuges, parks, and other areas relatively undisturbed by humans, the chances of finding significant evolutionary changes there seem more remote than in what amount to laboratories of evolution where both the environment and the species mix are in a state of flux.

Apart from the biological impact of the human population explosion, what can be said about human evolution? By the usual biological criteria for species, *Homo sapiens* constitutes a single polytypic species. According to some writers, biological evolution has virtually ceased in humans because cultural advances have been so rapid that cultural evolution has superseded biological evolution. The truth is, however, that we know very little about the present course of biological evolution in humans. The transition from a hunting-gathering nomadic culture to agriculture occurred only about 10,000 years or some 400 human generations ago. The control of infectious diseases as major causes of death has become possible just during the past century, and the treatment of hereditary diseases is even more recent. It seems inescapable that these events have changed the selection pressures operating in human populations. Similarly, the greatly increased mobility of the human population has undoubtedly enhanced the amount of gene flow among human populations and diminished the importance of random genetic drift as a factor in human evolution. Furthermore, increased exposure to medical, industrial, and military mutagenic radiations and to chemical mutagens seems likely to have increased the mutation rates of human genes. The net effect of all these changes remains obscure, however, because even though *Homo sapiens* is undoubtedly the most carefully studied species of all, very few serious efforts have been made to study the ongoing evolution of the human species. Imagine, if you will, the reception given at the National Institutes of Health or the National Science Foundation to a research proposal to study the adaptive significance of hereditary differences in height or weight or IQ or, for that matter, to determine whether gentlemen in fact prefer blondes or ladies prefer their gentlemen tall, dark, and handsome. What might be done with such information if it became available poses an interesting question. Yet human evolution and population growth proceed apace, whether we know where we are headed or not.

Glossary

additive genetic variance. That portion of the total phenotypic variance due to genes with additive effects.

allele. One of an array of alternative forms of a gene, occupying the same locus on homologous chromosomes.

allometric. With reference to growth where the growth rate of one part of the body differs from that of other parts.

allopatric. Individuals living so far apart that they are not potential mates.

allozyme. One of two or more slightly different versions of the same enzyme.

amphimixis. Cross-fertilization.

anagenesis. Phyletic evolution; evolution along a single line of descent.

assortative mating. Nonrandom mating in which like tends to pair with like (positive) or unlike types pair (negative).

autecology. The ecology of a single organism or of a single species.

automixis. Self-fertilization.

balanced lethals. Lethal recessive genes so closely linked that crossing over is rare, the genes remain in repulsion, both homozygotes die, and only the heterozygotes survive.

balancing selection. Selection maintaining genetic polymorphism in a population.

biometry. The application of statistics to biological problems.

Bohr effect. Hemoglobin with a decreasing affinity for oxygen as the acidity increases.

ceteris paribus. Other things being equal.

chiasma (pl. chiasmata). A visible change in pairing affecting two nonsister chromatids out of the four chromatids in a tetrad in

233

the first meiotic prophase. The point of exchange of partners is the chiasma. Cytological manifestation of crossing over.

chromatids. Half-chromosomes joined by a single centromere resulting from longitudinal duplication of a chromosome, observable during prophase and metaphase, and becoming daughter chromosomes at anaphase.

chromosome. Nucleoprotein bodies in the nucleus, usually constant in number for any given species, and bearing the genes in linear order.

cladogenesis. Branching evolution, forming an evolutionary tree.

cline. A gradual change or geographic gradient in a continuous population in the frequencies of different phenotypes or genotypes.

cohort. A group of individuals of similar age.

conjugation. May refer to the pairing or union of chromosomes, nuclei, cells, or individuals.

crossing over. The reciprocal exchange of corresponding segments between two nonsister chromatids of homologous chromosomes.

deficiency (deletion). The absence of a segment from a chromosome, which may vary in size from a single nucleotide to a number of genes.

deme. A local breeding population or panmictic unit.

dioecious. With male and female gametes produced by different unisexual individuals.

directional selection. Selection toward a single optimum leading to a shift in gene frequencies or in the population mean of the character considered.

disruptive selection. Simultaneous selection for more than one optimum phenotype in a population leading to divergence toward more than one optimum and selection against the intermediate types. (Also called diversifying selection.)

DNA. Deoxyribonucleic acid; the genetic material.

dominant. Used with reference to both genes and characters. An allele is said to be dominant if its phenotypic effect is the same in both the homozygous and heterozygous condition. However, dominance may be complete, incomplete, or absent.

duplication. The occurrence of a chromosome segment more than once in the same chromosome or haploid genome.

ecological niche. All the relationships of an organism or a species to its physical and biological environment.

ecotype. An ecological race genetically adapted to a particular habitat as the result of natural selection.

ectotherm. A cold-blooded animal.

ED$_{50}$. The dose of pesticide effective in killing or knocking down 50 percent of the exposed individuals in a given time.

endemic. Native to or restricted to a particular locality or geographic region. Not introduced or naturalized.

epigean. Living on or above the surface of the ground.

epistasis. Suppression of the expression of a gene or genes by other genes not allelic to the genes suppressed. Similar to dominance but involving the interaction of nonallelic genes. Sometimes used to refer to all nonallelic interactions.

ethology. The study of the mechanisms and evolution of animal behavior.

fission. Asexual division of a unicellular organism into daughter cells.

gender. Refers to organisms where individuals are of opposite sex, either male or female.

gene flow. The spread of genes from one breeding population to others as the result of the migration of individuals.

gene pool. The sum total of the genes in a given breeding population at a particular time.

genetic load. The reduction in the average fitness of a population compared to the fitness of its optimum genotype, or the change in average fitness of a population associated with maintaining its variability. Major categories of load are the input load, the balanced load, and the substitutional load.

genome. The chromosome complement of a gamete or a complete haploid set of genes or chromosomes.

genotype. The entire genetic constitution of an organism.

group selection. Usually, selection invoked to explain the evolution of traits that seem to run contrary to the interests of the individual, but favor the group as a whole.

hard selection. Frequency- and density-independent selection.

heritability. In the usual or narrow sense, the ratio of the additive genetic variance to the total phenotypic variance. In the broad sense, that proportion of the total phenotypic variance that is genetic in origin.

hermaphrodite. An individual with both functional ovaries and testes.

heterogamy. Mating of dissimilar individuals.

heteroselection. Selection occurring in genetic systems closed because they are polymorphic for chromosomal rearrangements.

heterosis. Hybrid vigor. The superiority of heterozygous genotypes with respect to one or more traits in comparison with the corresponding homozygotes.

heterozygous. An individual having two different alleles at one or more gene loci. Also, having two different chromosome arrangements.

homeostasis. The tendency of a system to maintain a dynamic equilibrium and, in case of disturbance, to restore the equilibrium by its own regulatory mechanisms. Originally a physiological term, now developmental homeostasis in individuals and genetic homeostasis in populations have been identified.

homogamy. Mating of similar individuals.

homoselection. Selection occurring in open genetic systems where the chromosomes are homosequential.

homosequential. Chromosomes having the genes in the same linear order, i.e., lacking chromosomal rearrangements.

homozygous. Having identical alleles at corresponding loci on homologous chromosomes so that the organism breeds true with respect to these loci.

inbreeding. The mating of individuals more closely related than mates chosen at random from the population.

insertion. Transposition of a chromosome segment to a nonterminal location within a chromosome. Three breaks are required.

introgression. The incorporation of genes from one species into the gene pool of another species by hybridization and backcrossing.

inversion. Rotation of a chromosome segment through 180° so that the linear order of the genes is reversed relative to the rest of the chromosome.

isoalleles. Alleles so similar in their effects that special techniques are needed to distinguish between them.

isolating mechanism. Any genetically controlled factor that prevents or reduces interbreeding and gene flow between different populations.

karyotype. The somatic chromosomal complement of an individual or a related group of individuals.

linkage. The association of nonallelic genes in inheritance greater than expected from independent assortment owing to their location on the same chromosome. Genes borne on homologous chromosomes belong to the same linkage group.

locus (pl. loci). The position of a gene on a chromosome.

Ludwig effect. Genetic polymorphism maintained by diversifying selection in a heterogeneous environment.

macromutation. A mutation of major effect, or "systemic mutation," thought by Goldschmidt to be responsible for large-scale evolution.

meiosis. The reduction divisions during which the chromosome number is reduced from diploid to haploid. Two nuclear divisions during which the chromosomes divide only once.

meiotic drive. Any meiotic mechanism that results in unequal proportions of the two types of gametes produced by a heterozygote.

meristic variation. Variation in traits that can be enumerated such as fin rays in fish, bristles in flies, or tail vertebrae in mice.

meromixis. A genetic transfer in bacteria in which only a fraction of the genome of a donor cell is transferred to the recipient cell.

migration. In population genetics, transfer of genetic information among populations through movement of individuals from one population to another. Gene flow.

mitosis. The process by which the nucleus is divided into two daugher nuclei, each with a complement of genes and chromosomes identical to that of the original nucleus.

modifying factor. A gene that affects the expression of genes at other loci, often without other known effects.

monoecious. Producing male and female gametes in a single individual. Hermaphroditic.

mRNA. Messenger RNA; a population of large heterogeneous ribonucleic acid molecules formed by transcription from complementary DNA molecules and serving as a template for protein synthesis.

Muller's rachet. In asexual populations, the tendency to accumulate deleterious genes resulting from mutation.

multiple-factor hypothesis. The notion that quantitative traits result from the cumulative effects of arrays of genes at many loci, each locus having only a small effect.

mutation. In the broad sense, any sudden heritable change in the hereditary material not due to segregation or to genetic recombination. Thus, mutations may be genic, chromosomal, or extrachromosomal. In the narrow sense, gene mutations only.

mutation pressure. The continued, recurrent production of an allele by mutation, tending to increase its frequency in a population.

mutation rate. The frequency with which a particular mutation occurs per locus per generation.

mutator gene. A gene that influences the mutation rate of genes at other loci.

N_e. Effective size of a breeding population. An estimate of the number of individuals actually contributing genes to succeeding generations. Usually differs from and is smaller than the census number.

negative assortative mating. Nonrandom mating in which unlike types pair.

normal frequency curve. A symmetrical, bell-shaped curve often approximated when frequency distributions are plotted from observations on biological materials.

oligogene. A gene with a pronounced phenotypic effect. A major gene as opposed to polygenes, each of which has an individually small effect.

overdominance. Originally defined with reference to a single locus where $A_1A_1 <A_1A_2> A_2A_2$. Now sometimes used to refer to superiority of any heterozygote over its corresponding homozygotes.

pangenesis. Darwin's discredited theory of heredity, which involved pangenes coming from all parts of the body to form the gametes.

panmixia. A system of mating characterized by random choice of mating partners.

parapatric. Allopatric populations whose ranges are contiguous so that hybridization and gene flow are possible along the zone of contact.

parental investment. The "biological capital" invested by parents in their offspring.

parthenogenesis. The development of an individual from a germ cell (usually an egg) without fertilization.

peripatric. Populations living on the periphery of a species distribution. Thought to serve as founding populations in speciation.

phenotype. The sum total of the observable or measurable characteristics of an organism produced by its genotype interacting with the environment.

polygenes. Numerous genes of individually small effect controlling the expression of quantitative characters.

polymorphic. The existence of two or more discontinuous variants within a single breeding population.

polytypic. Refers to species composed of two or more genetically diverse populations, often called races or subspecies.

population structure. The way a species population is subdivided and distributed in space.

positive assortative mating. Nonrandom mating in which like types pair.

probit analysis. A statistical transformation widely used in biological assays where the response is quantal, i.e., all or none.

race. A subspecies or a geographical subdivision of a species. Partially isolated populations of the same species that differ in the frequencies of certain of the genes in their gene pools.

random genetic drift. The random fluctuations in gene frequencies owing to sampling errors, especially noticeable in effectively small populations.

random mating. Any individual of one sex having an equal probability of mating with any individual of the opposite sex.

recombination. The process by which progeny attain combinations of genes other than those found in their parents, the result of segregation, independent assortment, and crossing over.

reproduction. The process, sexual or asexual, by which organisms produce new individuals.

selection. Differential reproduction by different genotypes.

selective mating. Nonrandom mating in which one type of male or female mates more often in proportion to its numbers than other types.

self-sterility alleles. Systems of alleles that inhibit the growth on the stigma of any pollen grain carrying an allele similar to either of the alleles found in the female parent.

sex ratio. The relative proportion of males and females of a specified age distribution in a population. Often expressed as the number of males per 100 females.

sexual isolation. Reproductive isolation owing to a tendency toward positive assortative mating or homogamic mating.

sexual selection. Darwin's theory of differential mating success based on male competition or female mating "preference" or both.

soft selection. Frequency- and density-dependent selection.

stabilizing selection. Favors a single optimum in a population by the elimination of the more extreme variants. Also known as normalizing or centripetal selection.

stasipatric. Speciation, primarily chromosomal, occurring within the area occupied by the ancestral species.

stirps. Galton's postulated units of heredity, which are found in and determine the development of a fertilized egg; some pass to the body cells, others remain in the sexual cells.

stochastic. Random. Involving a random variable (a stochastic process) or involving chance or probability.

subspecies. Geographically distinct aggregates of breeding populations that differ genetically from one another and between which gene flow is restricted. (See **race.**)

sympatric. Coexisting in the same area, with the implication that interbreeding is at least possible.

transduction. Transfer of DNA, the hereditary material, from one bacterium to another via a bacteriophage.

transformation. Transfer of genetic information in bacteria where DNA from one strain is incorporated into the genome of another strain.

translocation. Change in location of a chromosome segment to another chromosome or to another part of the same chromosome. Reciprocal translocation involves an exchange of segments between two chromosomes.

tRNA. Transfer RNA; each kind of tRNA binds to a particular amino acid and transports it to the mRNA (messenger RNA) where it is incorporated into the protein being synthesized.

variance. A numerical estimate of the variability of phenotypic values about the arithmetic mean. The mean squared deviation from the mean.

Wahlund effect. When a species population is divided into a number of random-mating subpopulations, the proportion of homozygotes in the species population is increased as compared to the proportion present if mating were completely random throughout the species.

wild type. The most frequently observed or "normal" phenotype. Also the most frequent allele in natural populations.

Bibliography

Adams, J., E. D. Rothman, W. E. Kerr, and Z. L. Paulino. 1977. Estimation of the number of sex alleles and queen matings from diploid male frequencies in a population of *Apis mellifera*. Genetics 86: 583–96.

Anderson, E. 1949. Introgressive Hybridization. Wiley. New York.

Anonymous. 1973. Where have all the frogs gone? Mod. Med. 41(23): 20–24.

Barr, T. C. 1968. Cave ecology and the evolution of troglobites. Evol. Biol. 2: 35–102.

Bates, H. W. 1862. Contributions to an insect fauna of the Amazon Valley. Lepidoptera: Heliconidae. Trans. Linn. Soc. London. 23: 495–566.

Bateson, W. 1894. Materials for the Study of Variation. Macmillan. London.

Beach, F. A. 1950. The Snark was a Boojum. Amer. Psychol. 5: 115–24.

Bell, G. 1982. The Masterpiece of Nature: The Evolution and Genetics of Sexuality. University of California Press. Berkeley.

Bishop, J. A., and L. M. Cook. 1975. Moths, melanism, and clean air. Sci. Amer. 232 (1): 90–99.

Bodmer, W. F., and A. W. F. Edwards. 1960. Natural selection and the sex ratio. Ann. Human Genet. 24: 239–44.

Bouchard, T. J., D. R. Lykken, M. McGue, N. Segal, and A. Tellegen. 1990. Sources of human psychological differences: The Minnesota Study of Twins Reared Apart. Science 250: 223–28.

Brace, C. L. 1963. Structural reduction in evolution. Amer. Nat. 97: 39–49.

Brandon, R. N., and R. M. Burian. Eds. 1984. Genes, Organisms, Populations: Controversies over the Units of Selection. MIT Press. Cambridge, Mass.

Breckenridge, W. J. 1941. Amphibians and reptiles of Minnesota. Ph.D. thesis. University of Minnesota.

Browder, L. W. 1967. Pigmentation in *Rana pipiens*: A study in developmental genetics. Ph.D. thesis. University of Minnesota.

Browder, L. W. 1980. Developmental Biology. 2nd Ed. Saunders. Philadelphia.

Brower, L. P., F. H. Pough, and H. R. Meck. 1970. Theoretical investigations of automimicry. I. Single trial learning. Proc. Nat. Acad. Sci. 66: 1059–66.

Brues, A. M. 1964. The cost of evolution vs. the cost of not evolving. Evolution 18: 379–83.

Bull, J. J., and E. L. Charnov. 1988. How fundamental are Fisherian sex ratios? Oxford Surveys in Evol. Biol. 5: 96–135.

Cain, A. J. 1979. Introduction to general discussion. Proc. Royal Soc. London (B) 205: 599–604.

Campbell, B. Ed. 1972. Sexual Selection and the Descent of Man: 1871–1971. Aldine. Chicago.

Carpenter, G. H. D., and E. B. Ford. 1933. Mimicry. Methuen. London.

Carson, H. L. 1971. Speciation and the founder principle. Stadler Genet. Symp. 3: 51–70.

Carson, H. L. 1975. The genetics of speciation at the diploid level. Amer. Nat. 109: 83–92.

Carson, H. L. 1978. Speciation and sexual selection in Hawaiian Drosophila. In Ecological Genetics: The Interface. P. F. Brussard. Ed. Springer-Verlag. New York.

Carson, H. L. 1982. Speciation as a major reorganization of polygenic balances. In Mechanisms of Speciation. C. Barigozzi. Ed. Alan R. Liss. New York.

Carson, H. L., D. E. Hardy, H. T. Spieth, and W. S. Stone. 1970. The evolutionary biology of Hawaiian Drosophilidae. Evol. Biol. (Suppl.): 437–543.

Castle, W. E. 1903. The laws of heredity of Galton and Mendel, and some laws governing race improvement by selection. Proc. Amer. Acad. Arts Sci. 39: 223–42.

Charlesworth, D., and B. Charlesworth. 1976. Theoretical genetics of Batesian mimicry. II. Evolution of supergenes. J. Theor. Biol. 55: 305–24.

Charnov, E. L. 1975. Sex ratio selection in an age structured population. Evolution 29: 366–67.

Charnov, E. L. 1982. The Theory of Sex Allocation. Princeton University Press. Princeton, N.J.

Cherry, L. M., S. M. Case, and A. C. Wilson. 1978. Frog perspective on the morphological differences between humans and chimpanzees. Science 200: 209–11.

Chetverikov, S. S. [1926] 1961. On certain aspects of the evolutionary process from the standpoint of modern genetics. Zhurnal Exp. Biol. A2: 3–54. In Russian. (Proc. Amer. Phil. Soc. 105: 167–95. English translation.)

Clausen, J., and W. M. Hiesey. 1958. Experimental studies on the nature of species. IV. Genetic structure of ecological races. Carnegie Inst. Wash. Publ. 615: 1–312.

Clausen, J., D. D. Keck, and W. M. Hiesey. 1940. Experimental studies on the nature of species. I. The effect of varied environments on western North American plants. Carnegie Inst. Wash. Publ. 520: 1–452.

Collins, A. M., M. A. Brown, T. E. Rinderer, J. R. Harbo, and K. W. Tucker. 1987. Heritabilities of honey-bee alarm pheromone production. J. Heredity 78: 29–31.

Comstock, R. E. 1977. Quantitative genetics and the design of breeding programs. In Proceedings of the International Conference on Quantitative Genetics. E. Pollak, O. Kempthorne, and T. B. Bailey. Eds. Iowa State University Press. Ames.

Cook, L. M., R. R. Askew, and J. A. Bishop. 1970. Increasing frequency of the typical form of the Peppered Moth in Manchester. Nature 227: 1155.

Cook, L. M., G. S. Mani, and M. E. Varley. 1986. Postindustrial melanism in the Peppered Moth. Science 231: 611–13.

Cott, H. B. 1940. Adaptive Coloration in Animals. Methuen. London.

Crow, J. F. 1966. Evolution of resistance in hosts and pests. In Scientific Aspects of Pest Control. Publ. 1402. Nat. Acad. Sci.–Nat. Res. Council. Washington, D.C.

Crow, J. F. 1970. Genetic loads and the cost of natural selection. In Mathematical Topics in Population Genetics. K. Kojima. Ed. Springer-Verlag. New York.

Crow, J. F. 1986. Basic Concepts in Population, Quantitative, and Evolutionary Genetics. Freeman. New York.

Crow, J. F. 1988. The importance of recombination. In The Evolution of Sex. R. E. Michod and B. R. Levin. Eds. Sinauer. Sunderland, Mass.

Crow, J. F., and M. Kimura. 1970. An Introduction to Population Genetics Theory. Harper & Row. New York.

Dapkus, D. C. 1976. Differential survival involving the Burnsi phenotype in the northern leopard frog, *Rana pipiens*. Herpetologica 32: 325–27.

Dapkus, D., and D. J. Merrell. 1977. Chromosomal analysis of DDT-resistance in a long-term selected population of *Drosophila melanogaster*. Genetics 87: 685–97.

Darlington, C. D. 1937. Recent Advances in Cytology. 2nd Ed. Blakiston. Philadelphia.

Darlington, C. D. 1939. The Evolution of Genetic Systems. Cambridge University Press. Cambridge.

Darlington, C. D. 1956. Natural populations and the breakdown of classical genetics. Proc. Royal Soc. London (B) 145: 350–64.

Darwin, C. R. 1859. On the Origin of Species by means of Natural Selection. John Murray. London. (The Origin of Species by Means of Natural Selection. 6th Ed. 1872. Random House. New York.)

Darwin, C. R. 1871. The Descent of Man and Selection in Relation to Sex. John Murray. London.

Dawkins, R. 1982. The Extended Phenotype. Freeman. San Francisco.

Dawkins, R., and J. R. Krebs. 1979. Arms races within and between species. Proc. Royal Soc. London (B) 205: 489–511.

Depew, D. J., and B. H. Weber. Eds. 1985. Evolution at a Crossroads: The New Biology and the New Philosophy of Science. MIT Press. Cambridge, Mass.

De Vries, H. 1901–3. Die Mutationstheorie. Versuche und Beobachtungen über die Entstehung der Arten im Pflanzenreich. 2 vols. Veit. Leipzig. (Trans. J. B. Farmer and A. D. Darbishire. The Mutation Theory. Open Court. Chicago. 1909.)

Dobzhansky, T. 1937. Genetics and the Origin of Species. (2nd Ed. 1941; 3rd Ed. 1951.) Columbia University Press. New York.

Dobzhansky, T. 1940. Speciation as a stage of evolutionary divergence. Amer. Nat. 74: 312–21.

Dobzhansky, T. 1955. A review of some fundamental concepts and problems of population genetics. Cold Spring Harbor Symp. Quant. Biol. 20: 1–15.

Dobzhansky, T. 1968. On some fundamental concepts of Darwinian biology. Evol. Biol. 2: 1–34.

Dobzhansky, T. 1970. Genetics of the Evolutionary Process. Columbia University Press. New York.

Dobzhansky, T., F. J. Ayala, G. L. Stebbins, and J. W. Valentine. 1977. Evolution. W. H. Freeman. San Francisco.

Dobzhansky, T., B. Spassky, and N. Spassky. 1952. A comparative study of mutation rates in two ecologically diverse species of *Drosophila*. Genetics 37: 650–64.

Dobzhansky, T., B. Spassky, and N. Spassky. 1954. Rates of spontaneous mutation in the second chromosomes of the sibling species, *Drosophila pseudoobscura* and *Drosophila persimilis*. Genetics 39: 899–907.

Dougherty, E. C. 1955. Comparative evolution and the origin of sexuality. Syst. Zool. 4: 145–69.

Dunn, L. C. 1965. A Short History of Genetics. McGraw-Hill. New York.

East, E. M. 1910. A Mendelian interpretation of variation that is apparently continuous. Amer. Nat. 44: 65–82.

Edwards, A. W. F. 1962. Genetics and the human sex ratio. Adv. Genetics 11: 239–72.

Eldredge, H., and S. J. Gould. 1972. Punctuated equilibria: An alternative to phyletic gradualism. *In* Models of Paleobiology. T. J. M. Schopf. Ed. W. H. Freeman. San Francisco.

Elton, C. S. 1927. Animal Ecology. Sidgwick & Jackson. London.

Emlen, J. M. 1968a. A note on natural selection and the sex ratio. Amer. Nat. 102: 94–95.

Emlen, J. M. 1968b. Selection for the sex ratio. Amer. Nat. 102: 589–91.

Endler, J. A. 1986. Natural Selection in the Wild. Princeton University Press. Princeton, N.J.

Eshel, I. 1975. Selection on sex ratio and the evolution of sex determination. Heredity 34: 351–62.

Falconer, D. S. 1977. Why are mice the size they are? In Proceedings of the International Conference on Quantitative Genetics. E. Pollak, O. Kempthorne, and T. B. Bailey. Eds. Iowa State University Press. Ames.

Falconer, D. S. 1989. Introduction to Quantitative Genetics. 3rd Ed. Wiley. New York.

Felsenstein, J. 1976. The theoretical population genetics of variable selection and migration. Ann. Rev. Genet. 10: 263–80.

Fisher, R. A. 1918. The correlation between relatives on the supposition of Mendelian inheritance. Trans. Royal Soc. Edinburgh 52: 399–433.

Fisher, R. A. 1928a. The possible modification of the response of the wild type to recurrent mutations. Amer. Nat. 62: 115–26.

Fisher, R. A. 1928b. Two further notes on the origin of dominance. Amer. Nat. 62: 571–74.

Fisher, R. A. 1930. The Genetical Theory of Natural Selection. Clarendon. Oxford. (2nd Ed. Dover. New York. 1958.)

Fisher, R. A. 1931. The evolution of dominance. Biol. Rev. Cambridge Phil. Soc. 6: 345–68.

Fisher, R. A., and E. B. Ford. 1928. The variability of species in the Lepidoptera, with reference to abundance and sex. Trans. Ent. Soc. London 76: 367–84.

Flor, H. H. 1956. The complementary genic systems in flax and flax rust. Adv. Genetics 8: 29–54.

Ford, E. B. 1937. Problems of heredity in the Lepidoptera. Biol. Rev. Cambridge Phil. Soc. 12: 461–503.

Ford, E. B. 1940. Polymorphism and taxonomy. In The New Systematics. J. Huxley. Ed. Clarendon Press. Oxford.

Ford, E. B. 1953. The genetics of polymorphism in the Lepidoptera. Adv. Genetics 5: 43–87.

Ford, E. B. 1964. Ecological Genetics. Methuen. London.

Ford, E. B. 1975. Ecological Genetics. 4th Ed. Chapman and Hall. London.

Frydenberg, O. 1963. Population studies of a lethal mutant in Drosophila melanogaster. I. Behavior in populations with discrete generations. Hereditas 50: 89–116.

Futuyma, D. J. 1979. Evolutionary Biology. Sinauer. Sunderland, Mass.

Futuyma, D. J., and M. Slatkin. Eds. 1983. Coevolution. Sinauer. Sunderland, Mass.

Galton, F. 1889. Natural Inheritance. Macmillan. London.

Gardner, C. O. 1977. Quantitative genetic research in plants: Past accomplishments and research needs. In Proceedings of the International Conference on Quantitative Genetics. E. Pollak, O. Kempthorne, and T. B. Bailey. Eds. Iowa State University Press. Ames.

Gause, G. F. 1934. The Struggle for Existence. Williams & Wilkins. Baltimore.

Georghiou, G. P. 1969. Genetics of resistance to insecticides in house flies and mosquitoes. Exp. Parasitol. 26: 224–55.

Ghiselin, M. 1974. A radical solution to the species problem. Syst. Zool. 23: 536–44.

Ghiselin, M. T. 1974. The Economy of Nature and the Evolution of Sex. University of

California Press. Berkeley.

Gibbs, E. L., G. W. Nace, and M. B. Emmons. 1971. The live frog is almost dead. Bioscience 21: 1027–34.

Gillespie, J. H. 1991. The Causes of Molecular Evolution. Oxford University Press. New York.

Ginsburg, L. R. 1983. Theory of Natural Selection and Population Growth. Benjamin/Cummings. Menlo Park, Calif.

Glass, B. Ed. 1980. The Roving Naturalist: Travel Letters of Theodosius Dobzhansky. American Philosophical Society. Philadelphia.

Goldschmidt, R. 1940. The Material Basis of Evolution. (Reprint, 1982.) Yale University Press. New Haven, Conn.

Goldschmidt, R. 1945. Mimetic polymorphism, a controversial chapter of Darwinism. Quart. Rev. Biol. 20: 147–64; 205–30.

Gould, S. J. 1980. Is a new and general theory of evolution emerging? Paleobiology 6: 119–30.

Gould, S. J. 1982a. The uses of heresy: An introduction to R. Goldschmidt's "The Material Basis of Evolution." Yale University Press. New Haven, Conn.

Gould, S. J. 1982b. The meaning of punctuated equilibrium and its role in validating a hierarchical approach to macroevolution. In Perspectives on Evolution. R. Milkman. Ed. Sinauer. Sunderland, Mass.

Gould, S. J. 1983. The hardening of the modern synthesis. In Dimensions of Darwinism. M. Grene. Ed. Cambridge University Press. Cambridge.

Gould, S. J., and R. C. Lewontin. 1979. The spandrels of San Marco and the Panglossian paradigm: A critique of the adaptationist programme. Proc. Royal Soc. London (B) 205: 581–98.

Gould, S. J., and E. S. Vrba. 1982. Exaptation—a missing term in the science of form. Paleobiology 8: 4–15.

Grant, V. 1981. Plant Speciation. 2nd Ed. Columbia University Press. New York.

Gray, R. D. 1987. Faith and foraging: A critique of the paradigm argument from design. In Foraging Behaviour. A. C. Kamil, J. R. Krebs, and H. R. Pulliam. Eds. Plenum. New York.

Gray, R. D. 1988. Metaphors and methods: Behavioural ecology, panbiogeography and the evolving synthesis. In Evolutionary Processes and Metaphors. M.-W. Ho and S. W. Fox. Eds. Wiley. New York.

Grinnell, J. 1914. An account of the mammals and birds of the Lower Colorado Valley. Univ. Calif. Publ. Zool. 12: 51–294.

Hailman, J. P. 1988. Operationalism, optimality and optimism: Suitabilities versus adaptations of organisms. In Evolutionary Processes and Metaphors. M.-W. Ho and S. W. Fox. Eds. Wiley. New York.

Haldane, J. B. S. 1930. A note on Fisher's theory of the origin of dominance and on a correlation between dominance and linkage. Amer. Nat. 64: 87–90.

Haldane, J. B. S. 1932. The Causes of Evolution. Longmans, Green. London. (Reprint. Cornell University Press. Ithaca, N.Y. 1966.)

Haldane, J. B. S. 1939. The theory of the evolution of dominance. J. Genetics 37: 365–74.

Haldane, J. B. S. 1954. The origins of life. New Biology 16: 12–27.

Haldane, J. B. S. 1957. The cost of natural selection. J. Genet. 55: 511–24.

Haldane, J. B. S. 1960. More precise expressions for the cost of natural selection. J. Genet. 57: 351–60.

Haldane, J. B. S. 1964. A defense of beanbag genetics. Perspect. Biol. Med. 7: 343–59.

Hamilton, W. D. 1964. The genetical evolution of social behavior. J. Theor. Biol. 7: 1–16;

17–52.

Hamilton, W. D. 1967. Extraordinary sex ratios. Science 156: 477–88.

Hardy, G. H. 1908. Mendelian proportions in a mixed population. Science 28: 49–50.

Hartl, D. 1980. Principles of Population Genetics. Sinauer. Sunderland, Mass.

Hartl, D. L., and S. W. Brown. 1970. The origin of male haploid genetic systems and their expected sex ratio. Theor. Pop. Biol. 1: 185–90.

Hedrick, P. W., M. E. Ginevan, and E. P. Ewing. 1976. Genetic polymorphism in heterogeneous environments. Ann. Rev. Ecol. Syst. 7: 1–32.

Henderson, L. J. 1913. The Fitness of the Environment. Macmillan. New York.

Heuts, M. J. 1953. Regressive evolution in cave animals. Symp. Soc. Exptl. Biol. 7: 290–309.

Hexter, W. M. 1955. A population analysis of heterozygote frequencies in *Drosophila*. Genetics 40: 444–59.

Hirschfield, M. F., and D. W. Tinkle. 1975. Natural selection and the evolution of reproductive effort. Proc. Nat. Acad. Sci. 72: 2227–31.

Ho, M.-W., and S. W. Fox. 1988. Processes and metaphors in evolution. *In* Evolutionary Processes and Metaphors. M.-W. Ho and S. W. Fox. Eds. Wiley. New York.

Ho, M.-W., and P. T. Saunders. Eds. 1984. Beyond Neo-Darwinism: Introduction to the New Evolutionary Paradigm. Academic Press. London.

Hubby, J. L., and R. C. Lewontin. 1966. A molecular approach to the study of genic heterozygosity in natural populations. I. The number of alleles at different loci in *Drosophila pseudoobscura*. Genetics 54: 577–94.

Hull, D. 1976. Are species really individuals? Syst. Zool. 25: 174–91.

Hull, D. 1978. A matter of individuality. Phil. Sci. 45: 335–60.

Hutchinson, G. E. 1957. Concluding remarks. Cold Spring Harbor Symp. Quant. Biol. 22: 415–27.

Hutchinson, G. E. 1978. An Introduction to Population Ecology. Yale University Press. New Haven, Conn.

Huxley, J. S. 1932. Problems of Relative Growth. Methuen. London.

Huxley, J. Ed. 1940. The New Systematics. Clarendon Press, Oxford.

Huxley, J. 1943. Evolution: The Modern Synthesis. Harper. New York.

Huxley, J. 1954. The evolutionary process. *In* Evolution as a Process. J. Huxley, A. C. Hardy, and E. B. Ford. Eds. Allen & Unwin. London.

James, J. W. 1965. Simultaneous selection for dominant and recessive mutants. Heredity 20: 142–44.

Jepsen, G. L., E. Mayr, and G. G. Simpson. Eds. 1949. Genetics, Paleontology, and Evolution. Princeton University Press, Princeton, N.J.

Johannsen, W. 1903. Über Erblichkeit in Populationen und in reinen Linien. Fischer. Jena.

Johannsen, W. 1909. Elemente der exakten Erblichkeitslehre. Fischer. Jena.

Johnson, C. 1966. Species recognition in the *Hyla versicolor* complex. Texas J. Sci. 18: 361–64.

Kamil, A. C., J. R. Krebs, and H. R. Pulliam. Eds. 1987. Foraging Behaviour. Plenum. New York.

Kamil, A. C., and T. D. Sargent. Eds. 1981. Foraging Behavior. Ecological, Ethological, and Psychological Approaches. Garland. New York.

Kempthorne, O. 1977. Introduction. Proceedings of the International Conference on Quantitative Genetics. Iowa State University Press. Ames.

Kettlewell, H. B. D. 1973. The Evolution of Melanism. Clarendon Press. Oxford.

Kimura, M. 1955. Stochastic processes and distribution of gene frequencies under natural selection. Cold Spring Harbor Symp. Quant. Biol. 20: 33–53.

Kimura, M. 1960. Optimum mutation rate and degree of dominance as determined by the principle of minimum genetic load. J. Genetics 57: 21–34.

Kimura, M. 1968. Evolutionary rate at the molecular level. Nature 217: 624–26.

Kimura, M. 1983. The Neutral Theory of Molecular Evolution. Cambridge University Press. Cambridge.

Kimura, M., and T. Ohta. 1971. Protein polymorphism as a phase of molecular evolution. Nature 229: 407–69.

King, J. L. 1967. Continuously distributed factors affecting fitness. Genetics 55: 483–92.

King, J. L., and T. H. Jukes. 1969. Non-Darwinian evolution: Random fixation of selectively neutral mutations. Science 164: 788–98.

King, M.-C., and A. C. Wilson. 1975. Evolution at two levels in humans and chimpanzees. Science 188: 107–16.

Kolata, G. B. 1975. Evolution in DNA: Changes in gene regulation. Science 189: 446–47.

Lack, D. L. 1943. The Life of the Robin. Penguin. Harmondsworth.

Lack, D. L. 1954. The Natural Regulation of Animal Numbers. Clarendon Press. Oxford.

Lack, D. L. 1966. Population Studies of Birds. Clarendon Press. Oxford.

Laidlaw, H. H., and R. E. Page. 1984. Polyandry in honey bees (Apis mellifera L.): Sperm utilization and intracolony genetic relationships. Genetics 108: 985–97.

Lamotte, M. 1951. Recherches sur la structure génétique des populations de Cepaea nemoralis L. Bull. Biol. France, Suppl. 35: 1–239.

Lamotte, M. 1959. Polymorphism of natural populations of Cepaea nemoralis. Cold Spring Harbor Symp. Quant. Biol. 24: 65–84.

Lande, R. 1980. Genetic variation and phenotypic evolution during allopatric speciation. Amer. Nat. 116: 463–79.

Lande, R. 1981. The minimum number of genes contributing to quantitative variation between and within populations. Genetics 99: 541–53.

Lande, R. 1983. The response to selection on major and minor mutations affecting a metrical trait. Heredity 50: 47–65.

Larson, A., E. M. Prager, and A. C. Wilson. 1984. Chromosomal evolution, speciation, and morphological change in vertebrates: The role of social behaviour. Chromosomes Today 8: 215–28.

Levene, H. 1953. Genetic equilibrium when more than one ecological niche is available. Amer. Nat. 87: 331–33.

Levins, R., and R. Lewontin. 1985. The Dialectical Biologist. Harvard University Press. Cambridge, Mass.

Lewis, H. 1953. The mechanism of evolution in the genus Clarkia. Evolution 7: 1–20.

Lewis, H. 1962. Catastrophic selection as a factor in speciation. Evolution 16: 257–71.

Lewis, H. 1966. Speciation in flowering plants. Science 152: 167–72.

Lewontin, R. C. 1968. Introduction. In Population Biology and Evolution. R. C. Lewontin. Ed. Syracuse University Press. Syracuse, N.Y.

Lewontin, R. C. 1970. The units of selection. Ann. Rev. Ecol. Syst. 1: 1–18.

Lewontin, R. C. 1974. The Genetic Basis of Evolutionary Change. Columbia University Press. New York.

Lewontin, R. C. 1979a. Fitness, survival, and optimality. In Analysis of Ecological Systems. D. J. Horn, G. R. Stairs, and R. D. Mitchell. Eds. Ohio State University Press. Columbus.

Lewontin, R. C. 1979b. Sociobiology as an adaptationist program. Behavioral Sci. 24: 5–14.

Lewontin, R. C. 1982. Human Diversity. Freeman. San Francisco.

Lewontin, R. C., and J. L. Hubby. 1966. A molecular approach to the study of genic heterozygosity in natural populations. II. Amount of variation and degree of heterozygosity in natural populations of Drosophila pseudoobscura. Genetics 54: 595–609.

L'Héritier, P., and G. Teissier. 1934. Une expérience de sélection naturelle. Courbe d'élimination du gène "Bar" dans une population de Drosophiles en équilibre. C. R. Soc. Biol. 117: 1048–51.

Li, C. C. 1948. An Introduction to Population Genetics. National Peking University Press. Peiping.

Lotka, A. J. 1925. Elements of Physical Biology. Williams & Wilkins. Baltimore.

Lowenstein, J. M. 1985. Molecular approaches to the identification of species. Amer. Sci. 73: 541–47.

McKusick, V. A. 1988. Mendelian Inheritance in Man. 8th Ed. Johns Hopkins University Press. Baltimore, Md.

Malécot, G. 1948. Les Mathématiques de l'Hérédité. Masson. Paris.

Mangelsdorf, A. J. 1952. Gene interaction in heterosis. In Heterosis. J. W. Gowen. Ed. Iowa State College Press. Ames.

Mather, K. 1941. Variation and selection of polygenic characters. J. Genet. 41: 159–93.

Maugh, T. H. 1972. Frog shortage possible this winter. Science 178: 387.

Maynard Smith, J. 1971. The origin and maintenance of sex. In Group Selection. G. C. Williams. Ed. Aldine/Atherton. Chicago.

Maynard Smith, J. 1978. The Evolution of Sex. Cambridge University Press. Cambridge.

Maynard Smith, J., and G. R. Price. 1973. The logic of animal conflict. Nature 246: 15–18.

Mayr, E. 1942. Systematics and the Origin of Species. Columbia University Press. New York.

Mayr, E. 1954. Change of genetic environment and evolution. In Evolution as a Process. J. Huxley, A. C. Hardy, and E. B. Ford. Eds. Allen and Unwin. London.

Mayr, E. 1959. Where are we? Genetics and twentieth century Darwinism. Cold Spring Harbor Symp. Quant. Biol. 24: 1–14.

Mayr, E. 1963. Animal Species and Evolution. Harvard University Press. Cambridge, Mass.

Mayr, E. 1970. Populations, Species, and Evolution. Harvard University Press. Cambridge, Mass.

Mayr, E. 1982. The Growth of Biological Thought. Harvard University Press. Cambridge, Mass.

Mayr, E. 1988. Toward a New Philosophy of Biology. Harvard University Press. Cambridge, Mass.

Mayr, E. 1991. One Long Argument: Charles Darwin and the Genesis of Modern Evolutionary Thought. Harvard University Press. Cambridge, Mass.

Mayr, E., and W. B. Provine. Ed. 1980. The Evolutionary Synthesis. Harvard University Press. Cambridge, Mass.

Merrell, D. J. 1951. Interspecific competition between Drosophila melanogaster and Drosophila funebris. Amer. Nat. 85: 159–69.

Merrell, D. J. 1953a. Gene frequency changes in small laboratory populations of Drosophila melanogaster. Evolution 7: 95–101.

Merrell, D. J. 1953b. Selective mating as a cause of gene frequency changes in laboratory populations of Drosophila melanogaster. Evolution 7: 287–96.

Merrell, D. J. 1965. Competition involving dominant mutants in experimental populations of *Drosophila melanogaster*. Genetics 52: 165–89.

Merrell, D. J. 1968. A comparison of the estimated size and the "effective size" of breeding populations of the leopard frog, *Rana pipiens*. Evolution 22: 274–83.

Merrell, D. J. 1969a. Natural selection in a leopard frog population. J. Minn. Acad. Sci. 35: 86–89.

Merrell, D. J. 1969b. The evolutionary role of dominant genes. *In* Genetics Lectures. Vol. 1: 167–94. Oregon State University Press. Corvallis.

Merrell, D. J. 1969c. Limits on heterozygous advantage as an explanation of polymorphism. J. Heredity 60: 180–82.

Merrell, D. J. 1972. Laboratory studies bearing on pigment pattern polymorphisms in wild populations of *Rana pipiens*. Genetics 70: 141–61.

Merrell, D. J. 1975. In defense of frogs. Science 189: 838.

Merrell, D. J. 1977. Life history of the leopard frog, *Rana pipiens*, in Minnesota. Occ. Papers Bell Mus. Nat. Hist. Univ. Minn. 15: 1–23.

Merrell, D. J. 1981. Ecological Genetics. University of Minnesota Press. Minneapolis.

Merrell, D. J., and C. F. Rodell. 1968. Seasonal selection in the leopard frog, *Rana pipiens*. Evolution 22: 284–88.

Merrell, D. J., and J. C. Underhill. 1956a. Competition between mutants in experimental populations of *Drosophila melanogaster*. Genetics 41: 469–85.

Merrell, D. J., and J. C. Underhill. 1956b. Selection for DDT resistance in inbred, laboratory, and wild stocks of *Drosophila melanogaster*. J. Econ. Ent. 49: 300–306.

Michod, R. E., and B. R. Levin. Eds. 1988. The Evolution of Sex. Sinauer. Sunderland, Mass.

Milkman, R. D. 1967. Heterosis as a major cause of heterozygosity in nature. Genetics 55: 493–95.

Milkman, R. D. 1982. Toward a unified selection theory. *In* Perspectives on Evolution. R. D. Milkman. Ed. Sinauer. Sunderland, Mass.

Moll, R. H., M. F. Lindsey, and H. F. Robinson. 1964. Estimates of genetic variances and levels of dominance in maize. Genetics 49: 411–23.

Moore, J. A. 1942. An embryological and genetical study of *Rana burnsi* Weed. Genetics 27: 408–16.

Mukai, T., S. Z. Chigusa, L. E. Mettler, and J. F. Crow. 1972. Mutation rate and dominance of genes affecting viability in *Drosophila melanogaster*. Genetics 72: 335–55.

Müller, F. 1878. Notes on Brazilian entomology. Trans. Ent. Soc. London. Pt. 3: 211–23.

Muller, H. J. 1925. Why polyploidy is rarer in animals than in plants. Amer. Nat. 59: 346–53.

Muller, H. J. 1932a. Some genetic aspects of sex. Amer. Nat. 66: 118–38.

Muller, H. J. 1932b. Further studies on the nature and causes of gene mutations. Proc. 6th Internat. Cong. Genetics 1: 213–55.

Muller, H. J. 1940. Bearings of the "Drosophila" work on systematics. *In* The New Systematics. J. Huxley. Ed. Clarendon Press. Oxford.

Muller, H. J. 1942. Isolating mechanisms, evolution, and temperature. Biol. Symp. 6: 71–125.

Muller, H. J. 1948. Evidence of the precision of genetic adaptation. Harvey Lectures 43: 165–229.

Muller, H. J. 1949. Redintegration of the symposium on genetics, paleontology, and evolution. *In* Genetics, Paleontology, and Evolution. G. L. Jepsen, E. Mayr, and G. G. Simpson. Eds. Princeton University Press. Princeton, N.J.

Muller, H. J. 1950. Our load of mutations. Amer. J. Human Genet. 2: 111–76.

Nei, M. 1987. Molecular Evolutionary Genetics. Columbia University Press. New York.

Nevo, E. 1978. Genetic variation in natural populations: Patterns and theory. Theor. Pop. Biol. 13: 121–77.

Nevo, E. 1988. Genetic diversity in nature: Patterns and theory. Evol. Biol. 23: 217–47.

Nevo, E. 1991. Evolutionary theory and processes of active speciation and adaptive radiation in subterranean mole rats, *Spalax ehrenbergi* superspecies, in Israel. Evol. Biol. 25: 1–125.

Nevo, E., A. Beiles, and R. Ben Shlomo. 1984. The evolutionary significance of genetic diversity: Ecological, demographic and life history correlates. *In* Evolutionary Dynamics of Genetic Diversity. G. S. Mani. Ed. Springer-Verlag. Berlin.

Nicholson, A. J. 1927. A new theory of mimicry in insects. Austral. Zool. 5: 10–104.

Nilsson-Ehle, H. 1909. Kreuzungsuntersuchungen an Hafer und Weisen. Lunds. Univ. Aarsk. N. F. 5: 1–122.

Nitecki, M. H. Ed. 1983. Coevolution. University of Chicago Press. Chicago.

Nordenskiöld, E. 1928. The History of Biology. Knopf. New York. (Tr. from 1920–24 Swedish edition.)

Nuttall, G. H. F. 1904. Blood Immunity and Blood Relationship. Cambridge University Press. New York.

O'Donald, P. 1980. Genetic Models of Sexual Selection. Cambridge University Press. Cambridge.

Oyama, S. 1988. Stasis, development and heredity. *In* Evolutionary Processes and Metaphors. M.-W. Ho and S. W. Fox. Eds. Wiley. New York.

Patterson, C. 1978. Evolution. Cornell University Press. Ithaca, N.Y.

Pearl, R. 1925. The Biology of Population Growth. Knopf. New York.

Pearson, K. 1904. On a generalized theory of alternative inheritance, with special reference to Mendel's laws. Phil. Trans. Royal Soc. London (A) 203: 53–86.

Peterson, J. R., and D. J. Merrell. 1983. Rare male mating disadvantage in *Drosophila melanogaster*. Evolution 37: 1306–16.

Plunkett, C. R. 1932. Temperature as a tool of research in phenogenetics: Methods and results. Proc. 6th Internat. Cong. Genetics 1: 158–60.

Plunkett, C. R. 1933. A contribution to the theory of dominance. Amer. Nat. 67: 84–85.

Poulton, E. B. 1912. Darwin and Bergson on the interpretation of evolution. Bedrock 1: 48–65.

Price, P. W. 1980. Evolutionary Biology of Parasites. Princeton University Press. Princeton, N.J.

Prout, T. 1964. Observations on structural reduction in evolution. Amer. Nat. 98: 239–49.

Provine, W. B. 1971. The Origins of Theoretical Population Genetics. University of Chicago Press. Chicago.

Provine, W. B. 1978. The role of mathematical population geneticists in the evolutionary synthesis of the 1930s and 1940s. Stud. Hist. Biol. 2: 167–92.

Provine, W. B. 1986. Sewall Wright and Evolutionary Biology. University of Chicago Press. Chicago.

Punnett, R. C. 1915. Mimicry in Butterflies. Cambridge University Press. Cambridge.

Ralin, D. B., M. A. Romano, and C. W. Kilpatrick. 1983. The tetraploid tree frog *Hyla versicolor*: Evidence for a single origin from the diploid *H. chrysoscelis*. Herpetologica 39: 212–25.

Rendel, J. M. 1959. Canalization of the scute phenotype of *Drosophila*. Evolution 13: 425–39.

Rendel, J. M. 1962. Evolution of dominance. *In* The Evolution of Living Organisms. G. W. Leeper. Ed. Melbourne University Press. Melbourne.

Rensch, B. 1960. Evolution above the Species Level. Columbia University Press. New York.

Rieger, R., A. Michaelis, and M. M. Green. 1976. Glossary of Genetics and Cytogenetics. 4th Ed. Springer-Verlag. Berlin.

Sage, R. D., P. V. Loiselle, P. Basasibwaki, and A. C. Wilson. 1984. Molecular versus morphological change among cichlid fishes of Lake Victoria. *In* Evolution of Fish Species Flocks. A. A. Echelle and I. Kornfield. Eds. University of Maine Press. Orono.

Sang, J. H. 1984. Genetics and Development. Longman. London.

Saunders, P. T. 1988. Sociobiology: A house built on sand. *In* Evolutionary Processes and Metaphors. M.-W. Ho and S. W. Fox. Eds. Wiley. New York.

Schaller, G. B. 1972. The Serengeti Lion. University of Chicago Press. Chicago.

Schmalhausen, I. I. 1949. Factors of Evolution. Blakiston. Philadelphia.

Schmidt, G. D., and L. S. Roberts. 1985. Foundations of Parasitology. 3rd Ed. Times Mirror/Mosby. St. Louis, Mo.

Sheppard, P. M. 1953. Polymorphism and population studies. Symp. Soc. Exp. Biol. 7: 274–89.

Sheppard, P. M. 1954. Evolution in bisexually reproducing organisms. *In* Evolution as a Process. J. Huxley, A. C. Hardy, and E. B. Ford. Eds. Allen & Unwin. London.

Sheppard, P. M. 1961. Some contributions to population genetics resulting from the study of the Lepidoptera. Adv. Genetics 10: 165–216.

Sheppard, P. M. 1975. Natural Selection and Heredity. 4th Ed. Hutchinson. London.

Sheppard, P. M., and E. B. Ford. 1966. Natural selection and the evolution of dominance. Heredity 21: 139–47.

Simmons, M. J. 1976. Heterozygous effects of irradiated chromosomes on viability in *Drosophila melanogaster*. Genetics 84: 353–74.

Simmons, M. J., and J. F. Crow. 1977. Mutations affecting fitness in *Drosophila* populations. Ann. Rev. Genetics 11: 49–78.

Simpson, G. G. 1944. Tempo and Mode in Evolution. Columbia University Press. New York.

Simpson, G. G. 1950. The Meaning of Evolution. Yale University Press. New Haven, Conn.

Simpson, G. G. 1953. The Major Features of Evolution. Columbia University Press. New York.

Simpson, G. G. 1983. Fossils and the History of Life. Scientific American Books. New York.

Simpson, G. G., A. Roe, and R. C. Lewontin. 1960. Quantitative Zoology. Rev. Ed. Harcourt, Brace. New York.

Sleigh, M. A. 1973. The Biology of Protozoa. American Elsevier. New York.

Sober, E. 1984. The Nature of Selection. MIT Press. Cambridge, Mass.

Sokal, R. R., and P. H. A. Sneath. 1963. Principles of Numerical Taxonomy. Freeman. San Francisco.

Solbrig, O. T., and D. J. Solbrig. 1979. Introduction to Population Biology and Evolution. Addison-Wesley. Reading, Mass.

Spieth, H. T. 1982. Behavioral biology and evolution of the Hawaiian picture-winged species group of Drosophila. Evol. Biol. 14: 351–437.

Stanley, S. M. 1979. Macroevolution: Pattern and Process. Freeman. San Francisco.

Stansfield, W. D. 1977. The Science of Evolution. Macmillan. New York.

Stebbins, G. L. 1950. Variation and Evolution in Plants. Columbia University Press. New York.

Stebbins, G. L. 1960. The comparative evolution of genetic systems. *In* Evolution after

Darwin. I. The Evolution of Life. S. Tax. Ed. University of Chicago Press. Chicago.

Stebbins, G. L. 1971. Processes of Organic Evolution. 2nd Ed. Prentice-Hall. Englewood Cliffs, N.J.

Stern, C., and E. W. Schaeffer. 1943. On wild-type iso-alleles in *Drosophila melanogaster*. Proc. Nat. Acad. Sci. 29: 361–67.

Sturtevant, A. H. 1965. A History of Genetics. Harper & Row. New York.

Sved, J. A., T. E. Reed, and W. F. Bodmer. 1967. The number of balanced polymorphisms that can be maintained in a natural population. Genetics 55: 469–81.

Tauber, C. A., and M. J. Tauber. 1989. Sympatric speciation in insects. *In* Speciation and Its Consequences. D. Otte and J. A. Endler. Eds. Sinauer. Sunderland, Mass.

Templeton, A. R. 1982. Adaptation and the integration of evolutionary forces. *In* Perspectives on Evolution. R. Milkman. Ed. Sinauer. Sunderland, Mass.

Thoday, J. M. 1953. Components of fitness. Symp. Soc. Exptl. Biol. 7: 96–113.

Thompson, J. N. 1982. Interaction and Coevolution. Wiley. New York.

Thomson, K. S. 1988. Fisher's microscope, or the gradualist's dilemma. Amer. Sci. 76: 500–502.

Tomlinson, I. P. M. 1988. Major-gene models of sexual selection under cyclical natural selection. Evolution 42: 814–16.

Trivers, R. L. 1971. The evolution of reciprocal altruism. Quart. Rev. Biol. 46: 35–57.

Trivers, R. L. 1972. Parental investment and sexual selection. *In* Sexual Selection and the Descent of Man. B. Campbell. Ed. Aldine. Chicago.

Trivers, R. L. 1985. Social Evolution. Benjamin/Cummings. Menlo Park, Calif.

Turesson, G. 1922a. The species and the variety as ecological units. Hereditas 3: 100–113.

Turesson, G. 1922b. The genotypical response of the plant species to the habitat. Hereditas 3: 211–350.

Turner, J. R. G. 1967. Why does the genotype not congeal? Evolution 21: 645–56.

Turner, J. R. G. 1977. Butterfly mimicry: The genetical evolution of an adaptation. Evol. Biol. 10: 163–206.

Turner, J. R. G. 1983. "The hypothesis that explains mimetic resemblance explains evolution": The gradualist-saltationist schism. *In* Dimensions of Darwinism. M. Grene. Ed. Cambridge University Press. Cambridge.

Turrill, W. B. 1940. Experimental and synthetic plant taxonomy. *In* The New Systematics. J. Huxley. Ed. Clarendon Press. Oxford.

Underhill, J. C., and D. J. Merrell. 1966. Fecundity, fertility, and longevity of DDT-resistant and susceptible strains of *Drosophila melanogaster*. Ecology 47: 140–42.

Van Valen, L. 1973. A new evolutionary law. Evolutionary Theory 1: 1–30.

Volpe, E. P. 1955. A taxo-genetic analysis of the status of *Rana kandiyohi* Weed. Syst. Zool. 4: 75–82.

Volpe, E. P. 1960. Interaction of mutant genes in the leopard frog. J. Heredity 51: 150–55.

Volterra, V. 1926. Variazioni e fluttuazioni del numero d'individui in specie animali conviventi. Mem. Acad. Lincei. Ser. 6. Vol. 2: 31–113. (Variations and fluctuations of the number of individuals in animal species living together. Trans. in Animal Ecology. R. N. Chapman. McGraw-Hill, New York. 1931.)

Waddington, C. H. 1953. Epigenetics and evolution. Symp. Soc. Exp. Biol. 7: 186–99.

Waddington, C. H. 1957. The Strategy of the Genes. Allen & Unwin. London.

Waddington, C. H. 1968. The paradigm for the evolutionary process. *In* Population Biology and Evolution. R. C. Lewontin. Ed. Syracuse University Press. Syracuse, N.Y.

Wallace, B. 1981. Basic Population Genetics. Columbia University Press. New York.

Wallace, B. 1991. Fifty Years of Genetic Load: An Odyssey. Cornell University Press. Ithaca, N.Y.

Watson, J. B. 1930. Behaviorism. Norton. New York.

Weed, A. C. 1922. New frogs from Minnesota. Proc. Biol. Soc. Wash. 34: 107–10.

Weinberg, W. 1908. Über den Nachweis der Vererbung beim Menschen. Jahreshafte Verein f. vaterl. Naturkunde in Württemberg 64: 368–82.

White, M. J. D. 1945. Animal Cytology and Evolution. Cambridge University Press. Cambridge.

White, M. J. D. 1968. Models of speciation. Science 159: 1065–70.

White, M. J. D. 1978. Modes of Speciation. Freeman. San Francisco.

White, M. J. D., R. E. Blackith, R. M. Blackith, and J. Cheney. 1967. Cytogenetics of the *viatica* group of morabine grasshoppers. I. The "coastal" species. Austral. J. Zool. 15: 263–302.

Wickler, W. 1968. Mimicry in Plants and Animals. (R. D. Martin. Trans.). McGraw-Hill. New York.

Williams, G. C. 1966. Adaptation and Natural Selection. Princeton University Press. Princeton, N.J.

Williams, G. C. Ed. 1971. Group Selection. Aldine/Atherton. Chicago.

Williams, G. C. 1975. Sex and Evolution. Princeton University Press. Princeton. N.J.

Wills, C. 1981. Genetic Variability. Oxford University Press. New York.

Wilson, A. C. 1975. Evolutionary importance of gene regulation. Stadler Genetics Symp. 7: 117–34.

Wilson, A. C., G. L. Bush, S. M. Case, and M.-C. King. 1975. Social structuring of mammalian populations and rate of chromosomal evolution. Proc. Nat. Acad. Sci. 72: 5061–65.

Wilson, A. C., S. S. Carlson, and T. J. White. 1977. Biochemical evolution. Ann. Rev. Biochem. 46: 573–639.

Wilson, D. S. 1980. The Natural Selection of Populations and Communities. Benjamin/Cummings. Menlo Park, Calif.

Wilson, D. S. 1983. The group selection controversy: History and current status. Ann. Rev. Ecol. Syst. 14: 159–87.

Wilson, E. O. 1975. Sociobiology. Harvard University Press. Cambridge, Mass.

Wilson, E. O. 1978. On Human Nature. Harvard University Press. Cambridge, Mass.

Wilson, E. O., and W. H. Bossert. 1971. A Primer of Population Biology. Sinauer. Stamford, Conn.

Wright, S. 1929a. Fisher's theory of dominance. Amer. Nat. 63: 274–79.

Wright, S. 1929b. The evolution of dominance. Comment on Dr. Fisher's reply. Amer. Nat. 63: 556–61.

Wright, S. 1931. Evolution in Mendelian populations. Genetics 16: 97–159.

Wright, S. 1932. The roles of mutation, inbreeding, crossbreeding, and selection in evolution. Proc. 6th Internat. Congr. Genetics 1: 356–66.

Wright, S. 1934. Physiological and evolutionary theories of dominance. Amer. Nat. 68: 24–53.

Wright, S. 1940. The statistical consequences of Mendelian heredity in relation to speciation. *In* The New Systematics. J. Huxley. Ed. Clarendon Press. Oxford.

Wright, S. 1948. On the roles of directed and random changes in gene frequency in the genetics of populations. Evolution 2: 279–94.

Wright, S. 1949. Adaptation and selection. *In* Genetics, Paleontology, and Evolution. G. L. Jepsen, E. Mayr, and G. G. Simpson. Eds. Princeton University Press. Princeton, N.J.

Wright, S. 1960. Genetics and twentieth century Darwinism: A review and discussion. Amer. J. Human Genet. 12: 365–72.

Wright, S. 1964. Pleiotropy in the evolution of structural reduction and of dominance. Amer. Nat. 98: 65–69.

Wright, S. Evolution and the Genetics of Populations. University of Chicago Press. Chicago.

Vol. 1. 1968. Genetic and Biometric Foundations.

Vol. 2. 1969. The Theory of Gene Frequencies.

Vol. 3. 1977. Experimental Results and Evolutionary Deductions.

Vol. 4. 1978. Variability within and among Natural Populations.

Wynne-Edwards, V. C. 1962. Animal Dispersion in Relation to Social Behaviour. Oliver & Boyd. Edinburgh.

Index

David J. Merrell graduated from Rutgers University with a B.S. in biology in 1941. After his service in the U.S. Army during World War II, Harvard University awarded him a Ph.D. in biology with a concentration in genetics in 1948. Appointed to the faculty of the University of Minnesota in 1948, he served as professor of zoology, of genetics and cell biology, and of ecology, evolution, and behavior before becoming professor emeritus in 1987.